PERMAFROST

A Guide to Frozen Ground in Transition

PERMAFROST

A Guide to Frozen Ground in Transition

by Neil Davis

University of Alaska Press

Fairbanks, Alaska

Library of Congress Cataloging-in-Publication Data

Davis, T. Neil.
 Permafrost : a guide to frozen ground in transition / by Neil Davis.
  p. cm.
 Includes bibliographical references and index.
 ISBN 1-889963-19-4 (cloth : alk. paper)
 1. Frozen ground. I. Title.

 GB641 .D39 2000
 551.3'84--dc21   00-057719

International Standard Book Number: 1-889963-19-4
Library of Congress Catalog Number: 00-057719

Printed in Hong Kong by C&C Offset Printing Co., Ltd.

This publication was printed on acid-free paper that meets the minimum requirements for the American National Standard for Information Science—Permanence of Paper for Printed Library Materials ANSA Z39.48-1984.

Publication coordination by Deborah González, University of Alaska Press. Book design by Kim Arney. Cover concept and photograph by Neil Davis, production and prepress by Dixon J. Jones, Academic Media Services, University of Alaska Fairbanks.

To the memory of
Dr. Troy L. Péwé
1918–1999

Who made major contributions to the field of permafrost
through his scholarly work, his training of students, and his
giving generous help to colleagues.

The plaque reads:

Contents

Preface

Frozen ground is common underfoot: one-half of the land surface in the northern hemisphere freezes and thaws each year, and one-fifth of it contains perennially frozen ground. The consequences are both widespread and numerous, and they affect the lives of many people, especially those who reside or travel in the far north. Among the most obvious consequences of freeze and thaw processes are roller-coaster roadways and pavement breaks—many seem to develop almost as soon as builders complete new highways. These problems occur because northern roadways are alive with movement: they swell and shrink with each freeze and thaw, and as any underlying ***permafrost***[1] thaws, road foundations can sag destructively.

Irregular roadways are just one aspect of the northern environment attributable to ground permanently frozen or that freezes and thaws each year. Freeze and thaw processes help shape the surface of the land; they affect the vegetation and wildlife habitats. Very important from the human viewpoint are the problems freezing and thawing create for building foundations and engineering structures such as pipelines.

We all know that when the ground freezes it becomes more solid, and if the ground is saturated with water we might expect the ground to swell, by as much as 9%—simply because water expands its volume by 9% when it freezes. If this was all that was happening, the freezing and thawing of

---

1. Boldfaced, italicized terms indicate those contained in the Glossary, typically but not always at first mention in the text. A boldfaced figure reference indicates the first mention of the figure in the text, e.g., **Figure 1.1**. The book contains a section of color plates, each of which duplicates or nearly duplicates a gray-scale illustration within the text.

ground would have little consequence, and it would be a dull topic worthy of little discussion. But sometimes as it freezes, the ground expands by 50%, or even more. For this dramatic expansion to occur, some special process must be underway, and that process is the migration of water through the freezing *soil.* Water is an unusual substance with an abnormal set of properties that partially dictate water's rate of migration through and its interaction with soil. It is water's abnormal character that makes freezing and thawing soil so interesting and so relevant to life at middle and high latitudes. Taking it one step farther, that abnormal character is due to of the water molecule's strange shape and the forces involved in its interaction with soil.

Water's abnormality is important, and it has much affected the layout of this book. Following Chapter 1's brief introduction to permafrost and seasonally frozen ground, Chapter 2 contains a discussion of water's fundamental characteristics and the reasons for them. This discussion should give every reader, regardless of background, enough information that he or she can develop an understanding of and appreciation for the interesting processes occurring when the ground freezes and thaws. In Chapter 2, I also promote the concept of free energy and another concept that I have called The Universal System Happiness Rule. These concepts are valuable aids to understanding the ways of the world in general, and they are especially useful to developing insight into what happens when the ground freezes and thaws.

Acknowledgments

Dr. Kenneth R. Coates originally suggested the general form of this book. Dr. Robert B. Forbes, another member of the University of Alaska Press's Editorial Board, helped me plan the book, but of course I am totally responsible for its content. I am particularly grateful to Dr. Richard D. Reger for giving me guidance throughout the project, for providing photographs and for performing a thorough and highly valuable review of the manuscript. The manuscript received much revision after his reading and also after critical readings by Dr. Robert B. Forbes, Dr. Keith Van Cleve, and Dr. Carl S. Benson, Dr. Hugh M. French, plus readings by others, including Neal B. Brown and Dr. Troy L. Péwé. Dr. Péwé, who died just as this book went to press, also helped me by generously providing illustrations, documents and advice. Dr. Rudy J. Candler's helpful review of the early part of the book led to revisions to the discussion of chemical bonding. Dr. John S. Wettlaufer assisted much through discussion of the premelting phenomenon and reviewing drafts of material on that topic. I am grateful to University of Alaska's Geophysical Institute librarian Julia H. Triplehorn and Deirdre Helfferich of the University of Alaska Press for assistance in referencing. Rose Watabe helped me locate and reproduce high-altitude photographs archived at the Geophysical Institute, University of Alaska Fairbanks, and pilot Tom George gave assistance in obtaining aerial photographs in the Fairbanks area. My wife Rosemarie Davis read various versions of the manuscript and made many helpful suggestions to improve clarity. It has been a pleasure to work with Kim Arney who did the layout work on this book and helped me improve on the illustrations. I owe special thanks to Carla Helfferich for the application of her considerable editorial skills to improving the manuscript at several stages along the way, and to University of Alaska Press Manager Deborah Gonzalez who skillfully shepherded this project from start to finish.

A Short Introduction to Permafrost and Seasonally Frozen Ground

▪ Freeze, Froze, Frozen: What These Words Mean

In keeping with the general plan followed in this book—to start with the fundamentals and work up from there—a good place to begin is with what people mean when they say that the ground freezes or that it is frozen. Their statements tacitly assume that everyone understands that the word freezing in its various forms pertains to the conversion of water to ice, and that when the ground freezes it is the contained water's change of *state (of matter)* from liquid to solid that makes frozen ground different from that not frozen. Because water changes from the solid state to the liquid state at 0°C (32°F), this is a big number in the frozen ground business. Zero degrees centigrade is the melting temperature of ice, and it is also called the freezing point of water—despite the fact that water does not always turn to ice at this temperature, but more of that later.

▪ Frozen Ground in Turmoil

Like many other words, "frozen" has more than one meaning, so in addition to frozen ground we have things like frozen assets and people frozen in their tracks. Thus, frozen brings to mind both cold and immobility, but only the first really applies to frozen ground. Frozen ground definitely is mobile; it flows under load, it expands and contracts as its temperature changes, its soil particles shift and deform, and molecules of water move around between them. In certain circumstances, freezing ground can swell up like a balloon, and when it thaws it can shrink to a fraction of its frozen size or slither away down a slope. Furthermore, the movements that accompany freezing and thawing within the ground can alter the makeup of a soil by reorienting soil particles, compressing them, changing their shape, and sorting them according to size. So it is best not to think of freezing and frozen soil as a static entity. Albeit in slow motion, it churns and seethes; it is a thing alive that leads a most interesting life involving time scales ranging from hours to thousands of years.

What causes such lively behavior of freezing, frozen and thawing ground? In a word: water. Water gives frozen ground its special character and causes all the turmoil when the ground freezes or thaws. So after the following brief examination of the geography and consequences of permafrost and seasonally frozen ground, the road ahead leads to a close look at water in its own right in Chapter 2 and then, in Chapter 3, examination of what happens when the ground contains water and the temperature goes down below 0°C.

■ Definition of Permafrost

As noted in the Preface, this book contains a glossary that provides definitions of words italicized in the text, in most cases when they initially appear. *Permafrost* is an early example. A widely accepted definition of permafrost is that it is ground that has a temperature lower than 0°C (32°F) continuously, for at least two consecutive years. Much permafrost is thousands of years old, and some is of recent origin. Permafrost is forming now in places previously unfrozen: under thickening moss of boreal forests, and in the ground on the north side of new buildings (in the northern hemisphere) where the midday sun no longer shines. However, the twentieth century has been largely an era of global warming so, by and large, the amount of permafrost is in decline. Each winter the ground in many places freezes to depths of 1 or 2 meters but no permafrost forms because the following summer's warmth brings the temperature up to 0°C or above.

This definition of permafrost on the basis of the ground's temperature alone has a generality that may be useful in certain circumstances since it includes material such as rock that might contain no water at all. Nevertheless, my experience is that a person is hard pressed to avoid thinking of permafrost as anything other than soil that remains frozen year around; that is, to think of permafrost as a substance rather than a condition—and anyone who has tried to dig up permafrost knows that it is a tough substance indeed.

The word "permafrost" (permanent + frozen) also is a catch-all term used to describe the general area of science relating to all aspects of frozen or nearly frozen ground, past, present and future. *Geocryology* (earth + cold + study) also is a term for this field, and yet another in use is ***periglacial*** (near + glacier) processes.

■ Where Permafrost Exists

Permafrost is present beneath the surface of approximately one-fifth of the earth's land area (in the northern hemisphere, 22% of the land area) and ***subsea permafrost*** exists on the shallow continental shelves of cold polar seas. As **Figure 1.1** (bold face indicates the first reference to a figure or table) shows, permafrost is extensive at high latitudes, and much permafrost also exists in mountainous middle-latitude areas such as the Rocky Mountains in North America and the vast highlands of China and Tibet in Asia. Geocryologists call this perennially frozen ground ***alpine permafrost***, but the physical characteristics of this permafrost are no different than that in the continuous and discontinuous permafrost zones. Notice that the Eurasian permafrost extends southward to beyond the thirtieth parallel, and thus to the latitude of Florida and southern Texas, a clear demonstration that permafrost is not just a polar, nor even an arctic, phenomenon.[1]

At locations where the mean annual temperature of the air is near 0°C isolated pods of permafrost are found. They appear below the typically 1- to 2-m thick surface zone of annual freeze and thaw (called the ***active layer***) down to depths of 10 meters or so. Going north or up in altitude to cooler climes where the mean annual temperature is several degrees below 0°C, the active layer thins, and the layer containing permafrost thickens. Within this layer, perhaps 30 to 50 m thick near its southern fringe, permafrost is prevalent but discontinuous. In the northernmost parts of Alaska, Canada, and Siberia the active layer is very thin, perhaps only a few centimeters thick, and the permafrost layer in contact with it is continuous except near bodies of water that remain largely unfrozen. The permafrost can extend down to depths greater than 400 m; the deepest known is 1,450 m, at a location in Siberia.[2] See **Figure 1.2**.

---

1. Pointed out by Peter J. Williams, a reviewer of the manuscript.
2. Grave (1956), cited by Washburn (1980).

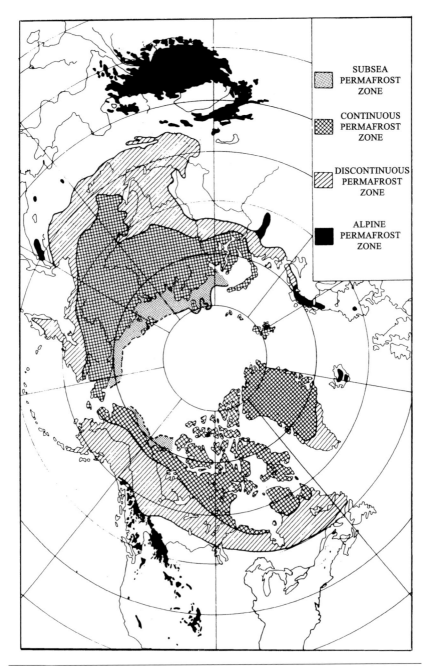

Figure 1.1 Permafrost in the northern hemisphere; After Péwé (1975a), modified slightly from his Figure 17.2.

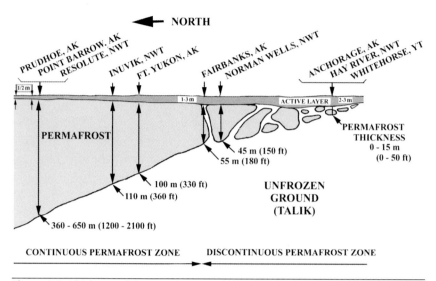

Figure 1.2 Thickness of the active layer and vertical distribution of permafrost in Alaska and western Canada. Based on Figure 4 by R. J. E. Brown (1975) with data added from Péwé (1982) and Washburn (1980).

Around the fringes of the Arctic Ocean is a region identified in Figure 1.1 as the subsea permafrost zone. The permafrost here is a relic of former times, having been formed beginning about 20,000 years ago or even earlier during times when sea level was some 90 meters lower than at present because so much water was locked up in glacial ice. The mean annual temperature at the sea floor off Alaska ranges from −0.7° to −3.4°C, so using the strict definition of permafrost as a condition of temperature requires that the material in the sea floor be called permafrost. However, because of the salinity of the nearby water and other factors, the upper 10 to 100 meters of the sea floor does not contain ice, and therefore is not frozen.[3] Deeper down, ice does bond the soil into a hard entity. The depth to the bottom of this bonded permafrost layer is not known.[4]

Notice in Figure 1.1 that the region of continuous permafrost extends outward from the pole to encompass the northern halves of Alaska and Canada, the northern two-thirds of Greenland and substantial parts of Siberia, Tibet, and China. The zone of discontinuous permafrost extends to

3. Osterkamp and Harrison (1976).
4. Rawlinson (1983) pp 4–8.

cover nearly all of Alaska (except low-altitude parts of the Aleutian Islands and southeast Alaska), most of Canada, parts of the Rocky Mountains in the western United States, all of Greenland, the central part of Iceland, northern Scandinavia, northern Europe, and large parts of Siberia, Tibet and China.

Because of the cyclical processes acting within it, the top layer of ground that undergoes seasonal freezing and thawing is particularly interesting. North American geocryologists tend to use the term "active layer" to describe the seasonal freeze-and-thaw layer in the continuous and discontinuous permafrost zones where it may be in contact with the top of the permafrost (called the **permafrost table**) or may overlie unfrozen ground. Russian and Chinese geocryologists apply a broader meaning to the term "active layer," allowing its definition to include all ground that seasonally freezes and thaws, both inside and outside the continuous and discontinuous permafrost zones. The Russian and Chinese usage makes the most sense because "active" is highly descriptive of the layer's mechanical processes that move water and soil to create (geologically speaking) rapid and profound changes occurring with or without underlying permafrost. Defined in the broad (Chinese and Russian) sense, the active layer covers much territory: in the northern hemisphere, 48% of the land area. As **Figure 1.3** shows, the layer covers virtually all of the mainland United States, and it dips down into the high country of Mexico. The thickness of the active layer (or what North American geocryologists call the **seasonally frozen layer**) is nearly zero close to the southern boundary of the United States, but it increases northward and with altitude to become near two meters in southern Canada. Farther north, underlying permafrost enforces a decrease in the thickness of the active layer, until, in the very far north, it declines to a few or few tens of centimeters. In the discontinuous permafrost zone layers of unfrozen soil, called **talik**, may lie between the permafrost table and the bottom of the active layer, as well as between blocks of frozen soil.

▪ Consequences of Permafrost and Seasonally Frozen Ground

1. Frozen ground is far less pervious to water than unfrozen ground, so by curtailing the downward seepage of water, permafrost modifies the environment for plants and animals, typically making that environment more wet in low areas—and even on hillsides, as, much to their chagrin, officials of the U.S. Bureau of Land Management (BLM) recently discovered. As a consequence of building a

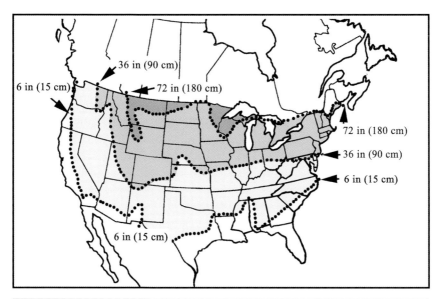

Figure 1.3 A highly generalized map based on one by Strock and Korel (1959) showing the approximate maximum depth of the active layer (the layer of seasonal freeze and thaw) in inches (cm). Because of local variations in climate, soil conductivity, and terrain, the actual depth of freezing at any location may differ greatly from that indicated on the map.

road across hillsides north of Fairbanks, Alaska, without benefit of permit to disturb land designated as wetlands, BLM was recently charged by the U.S. Corps of Engineers with breaking federal law. The corps believes that 48% of Alaska is federally designated wetland that cannot be disturbed without the corps' permission. The legal hassle arises because permafrost allows moss, sedges, bushes and trees normally found only in lowland areas to grow on upland areas, particularly those sloping to the north.[5]

2. Where the permafrost table comes close to the ground surface it forms an impenetrable barrier to roots. Plants and trees that grow atop this shallow permafrost have shallow root systems, and they tend to grow slowly because the soil is always cold.

3. Frost action, the process of freezing and thawing, changes the composition of the ground by altering soil particles and sorting

---

5. *Fairbanks Daily News-Miner* (1997).

them according to size, and it also can modify the very shape of the ground surface through erosion and downslope transport. Distinctive landforms are the result.

4. Events that cause permafrost to melt—such as climatic variations, forest fires, or human activities that destroy overlying ground cover—may create **thermokarst** topography, irregular terrain typically containing sinks and "drunken forests," as in **Figure 1.4**.

5. Permafrost literally puts paleozoologists, paleontologists, and archaeologists in direct contact with the ancient cold-climate plant, animal and human life of the **Pleistocene**, the recent geologic epoch that began 1.6 to 3 million years ago. Some permanently frozen ground, such as the extensive **loess** deposits in the vicinity of Fairbanks, Alaska, are rich in plant and animal remains preserved in or near original form. One famous example is Siberia's Beresovka mammoth, found in the early 1900s nearly intact and with flowers still in its mouth. Although it died 45,000 years ago, some of the mammoth's fleshy parts remained intact. Another important discovery made dur-

Figure 1.4 An island covered by dying trees sinks into a thermokarst lake near where the Alaska Highway crosses the Alaska-Canada border, in a region where warming during recent decades has caused shores to collapse as the permafrost below melts. See also Plate 1.

ing mining operations just north of Fairbanks, Alaska, in 1979 was of a nearly intact 36,000-year-old mummified bison now known as Blue Babe. Paleontologist R. Dale Guthrie of the University of Alaska Fairbanks in his book *Frozen Fauna of the Mammoth Steppe, the story of Blue Babe*[6] details the unearthing and study of the bison and provides a fascinating account of the setting and conditions that permitted preservation of the bison mummy, which is on display at the University of Alaska Museum in Fairbanks. (**Figure 1.5**)

6. Permafrost also provides indirect information about variations in climate during the Pleistocene because the existing distribution of temperature with depth depends in part on past mean annual air temperatures. By measuring the temperatures in holes drilled into thick permafrost, it is possible to gain some insight into air temperatures during past millennia.

7. A significant engineering problem associated with the freezing of the ground is *frost heave*. In places having fine-grained soils and plenty of water, frost heave can exert powerful forces that lift objects such as telephone poles and pilings up out of the ground.

8. Mainly because it can contain deposits of pure or nearly pure ice, permafrost is a serious geological hazard. As long as it remains frozen, permafrost typically is a tough material capable of supporting heavy loads (at least on short time scales), but when permafrost thaws, the melting of the ice can create voids in the ground and soupy mud flows. These ground failures destroy manmade structures such as roads, pipelines, homes, utility systems, and public buildings. Some examples:

   a. Thawing permafrost made a quagmire of parts of the Alaska (Alcan) Highway when if was first constructed in World War II, and in fact the word "permafrost" came into common use because of that experience (**Figure 1.6**);

   b. The permafrost underlying its 800-mile (1,300-km) route required that half of the trans-Alaska pipeline be built above ground, on expensive pilings that helped preserve the permafrost below (**Figure 1.7**);

   c. Within a few miles of the University of Alaska campus at Fairbanks—where they teach engineers and architects about the hazards of permafrost—many people have built homes over permafrost and then seen them become distorted or destroyed as the permafrost below melted (**Figure 1.8**).

---

6. Guthrie (1990).

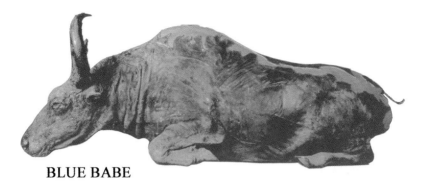

BLUE BABE

DIMA

Figure 1.5 Top: Blue Babe, a mummified bison that died 36,000 years ago, now resides in the University of Alaska Museum in Fairbanks.[a] (Photo: University of Alaska Museum). Bottom: Dima, a well-studied baby mammoth 7 to 8 months old when he died, was found in 1977 during gold mining operations in Siberia, north of Magadan. Dima, now displayed in the Leningrad Museum of Natural History, has reddish fur (mostly slipped off except around the feet) and is 144 cm high at the shoulder. He died more than 9,000 years ago, perhaps as long as 40,000 years ago.[b] Photograph courtesy of Dr. Kontrimavichus, Institute of Biological Problems of the North, Magadan, and the University of Alaska's Institute of Arctic Biology, Fairbanks, Alaska.

[a.] Guthrie, Mary Lee (1988); Guthrie, R. Dale (1990).
[b.] Guthrie, R. Dale (1990) pp 7–24.

Figure 1.6 At top a Caterpillar tractor tows a loaded sled and two trailers through the quagmire created by thawing permafrost during the construction of the Alcan Highway in 1942, and at bottom two soldiers face the daunting task of digging out a Cat unable to proceed on its own. University of Alaska Fairbanks Archives, Donavan B. Correll Collection.

Figure 1.7 A forest of refrigerated pilings supports the trans-Alaska pipeline where it snakes through the Alaska Range north of Isabel Pass. See also Plate 2.

■ The Mystery of Frozen Ground: Where Did All the Ice Come From, and How Did It Get There?

A long-standing challenge in the study of permafrost has been to explain the observation that frozen soil typically contains pods or layers of pure or nearly pure ice. Some of these may be only a millimeter thick while others may be several meters, as in **Figure 1.9**. Where did this pure ice come from, and exactly how did it get there?

The answer to this question is bound up in three general characteristics of water molecules: 1) in liquid form they tend to cling to each other more than most other kinds of molecules, 2) when they solidify into ice they cling to themselves so well that they tend to exclude molecules of other kinds, and 3) water molecules are capable of moving through unfrozen soil, frozen soil, and even ice, typically in the direction toward the coldest parts of these substances. Ordinary water is not an ordinary substance, and therein lies an interesting story.

Figure 1.8 Built over an ice wedge that melted, this house near the University of Alaska Fairbanks campus became so swaybacked that it had to be abandoned and destroyed. The collapse occurred quickly because this house was heated by hot water pipes placed in the concrete floor of the basement. The rows of white dots in the lower photograph mark the approximate location of the top of the house's foundation just before it was demolished. Top photo courtesy of Richard Reger.

Figure 1.9 A layer of ice 2 meters thick below ice wedges extending to the top of the permafrost. Photographed by Troy L. Péwé near Tuktoyaktuk on the Arctic seacoast.

The Whys and Wherefores of Permafrost

■ Water, Nature's Abundant Oddity

The most abundant single chemical substance near the surface of the earth is water,[1] and all living things are mostly water—somewhere between 60% and 90%. The average human being is 71%, and many of human society's measurements are in terms of water. Three very basic ones are:

1. The freezing and boiling points of water at sea level respectively define the 0° and 100° points on the centigrade (Celsius) temperature scale. (0°C is 273.15 K [the symbol K means degrees Kelvin].)
2. A liter is the volume of 1 kilogram of water. (Similarly, 1 gram is the mass or weight of 1 cubic centimeter of water.)
3. A *calorie* is the average amount of heat required to raise the temperature of 1 gram of water 1°C anywhere in the range 0°C to 100°C. To put it another way, the *specific heat* of water is 1 calorie.

The fact that water is plentiful does not by itself explain why life on this planet is based on water. It is of course convenient to have so much water around, but the reason why water is the crucial essence of life is the curiously odd nature of the water molecule. It has a strange shape and,

---

1. Specific information on the abundance of water near the earth's surface is in Appendix A.

partly for that reason, water possesses an abnormal set of properties. Some of these properties are largely responsible for the nature of frozen ground and the interesting changes that occur when the ground freezes and thaws.[2]

Water Molecules and the Forces that Hold Them Together, to Each Other, and to Other Substances Such as Soil

The water molecule and all other molecules are composed of atoms bonded together, and atoms are constructed from three fundamental building blocks: ***protons, neutrons***, and ***electrons***. The inner core of each atom, the nucleus, is composed of protons and neutrons each having about equal mass, but the neutron has no electrical charge, and the proton carries an elemental unit of positive charge. The electron carries the same amount of charge as the proton, but it is negative charge, and the electron has very little mass, 1/1,846 that of the proton.[3]

As portrayed in **Figure 2.1**, one proton and one electron joined together create the hydrogen atom, the lightest of all atoms. Two protons, two neutrons and two electrons joined create the next lightest atom (element), helium. All the other elements are similarly constituted; always with equal numbers of electrons and protons in each atom (but the number of neutrons can be equal, less or greater). The oxygen atom, an important component of the water molecule, has eight electrons swarming around a nucleus composed of eight protons and eight neutrons.

Atomic nuclei are held together by somewhat mysterious forces not of concern here because, while very strong, they are effective only on the tiny scale of atomic nuclei. Of much greater interest are the forces that bind the electron clouds of atoms to the nuclei and those that bind atoms together to form molecules. These are electromagnetic forces that depend on the electrical character of matter but not its mass. By contrast, the force of gravity depends on the mass character of matter but not the electric character. Gravity binds the earth to the sun, the moon to the earth, and it makes us

---

2. The overall discussion in this chapter relies heavily on three highly readable books: Davis and Day (1961); Knight (1967); Deming (1975). Another valuable source is the textbook by Pauling (1970), especially pp 420–46.
3. See Appendix A for more information on these particles and their interactions.

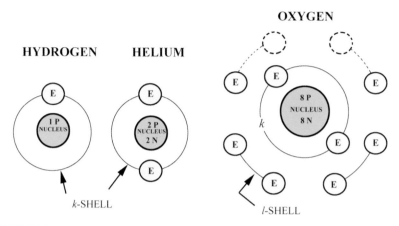

Figure 2.1 Atoms. Although shown here as fixed circular objects, the electrons (E) are smeared out to envelop the nucleus of each atom. The helium nucleus contains two protons (P) and two neutrons (N), and the oxygen nucleus has eight of each. The closed k shell of the helium atom makes it very stable and not prone to join with other atoms, but the outer shell of the oxygen atom (the l shell) contains two paired electrons plus two unpaired ones that easily form pairs with unpaired electrons attached to other atoms.

fall down when our feet slide out too far from beneath our center of mass. It is an inverse square force because the strength of the gravity force between two objects depends on the square of the distance (d in **Figure 2.2A**) between them as well as upon their masses m_1 and m_2. Thus if the distance between the two objects is doubled, the gravity force between them falls to one-fourth its former value. The gravity force acts along the line joining the centers of mass of two objects, and it always causes them to pull towards each other. Gravity causes only the most minor attraction between atomic nuclei and their surrounding swarms of electrons, an attraction so weak that it is inconsequential compared to electromagnetic force.

It is useful to think of electromagnetic force as having two parts. One part (the electro part) depends on the *position* of electrical charge, and the other part (the magnetic part) depends on the *motion* of electrical charge.

The part of electromagnetic force depending only on the position of charge is called the ***electrostatic force*** or the ***coulomb force***. As Figure 2.2A shows, the electrostatic force is like the force of gravity in that it acts only along the imaginary line drawn between the two objects involved and it grows weaker according to the square of their increasing separation. But since the electrostatic force depends on electrical charge instead of mass, and charge can be negative or positive, the force can be either attractive or

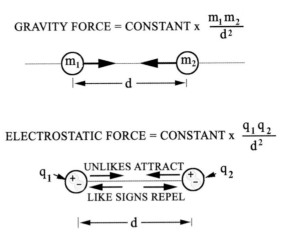

ELECTROSTATIC FORCE = CONSTANT x $\dfrac{q_1 q_2}{d^2}$

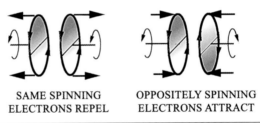

MAGNETIC FORCE BETWEEN SPINNING ELECTRONS

SAME SPINNING   OPPOSITELY SPINNING
ELECTRONS REPEL   ELECTRONS ATTRACT

Figure 2.2A   The gravity force (top) causes attraction between objects having mass. The electrostatic force (center) causes attraction between charges of opposite sign and repulsion between those of like sign. The magnetic force (bottom) generated by moving charge creates repulsion between electrons spinning in the same direction, and attraction if they spin in opposite directions.

repulsive. The electrostatic force between unlike charges q_1 and q_2 pulls them together and the force between charges of the same sign pushes them apart. It is the primary force holding the electrons in an atom to the atomic nucleus, and also, as will be seen shortly, one of the two powerful forces that bind atoms together to form molecules.

The magnetic part of electromagnetic force is created by moving electrical charge, and it acts only on moving charge. The electrons in an atom are orbiting around the atomic nucleus and they also spin like tops on their own axes. Such motion causes them to generate what is called a magnetic force (or magnetic field). Each spinning electron generates a magnetic force field with exactly the same shape as the magnetic field of a bar magnet, so like bar magnets, two spinning electrons brought close together will experience an attracting or repelling magnetic force, depending on how they are oriented. Furthermore, if the two electrons are spinning in not exactly opposite directions their magnetic force fields will interact to bring the spins into exact opposition, maximizing the magnetic force tending to bring the electrons together. As the two electrons approach each other they also experience a repulsive electrostatic force, but if the electrons spin in opposite directions the magnetic force field will overcome that repulsion. Then the two electrons will move toward each other until the repulsive electrostatic force just balances the attractive magnetic force.

This tendency for two electrons to join in this fashion is called electron pairing, and it is responsible for a very strong bonding between atoms, ***covalent bonding***. The simplest example of covalent bonding is the joining of two hydrogen atoms to form the hydrogen molecule H_2. Each atom consists of one electron attached to a proton nucleus, and when hydrogen atoms come close enough together with proper orientation the electrons pair up. They become linked in tandem like a pair of oxen constrained by a yoke representing the balance between the electrons' electrostatic (coulomb) repulsion and their mutual magnetic attraction. Like the oxen, the linked electrons go where they must, but always side by side, forever turning in unison to face the same direction while dragging their heavy nuclei tails behind, thereby creating rigid molecular structures. The paired electrons now behave as a unit, appearing to be mutually shared between the two nuclei, and so covalent bonding is also referred to as electron sharing.[4]

Cooperating with covalent bonding is another strong kind of chemical bonding, called ***ionic bonding*** (also electrostatic and valence bonding). It depends on the electrostatic (coulomb) attraction that occurs between two atoms when each distorts the electron swarm of the other. The negatively charged swarm (typically consisting mainly of one or more electron pairs) shifts toward one of the atoms, making it slightly negatively charged, and away from the other, leaving it somewhat positively charged. The ionic bond is the consequence—a mutual attraction of the distorted atoms, each of which appears to the other to be carrying electrical charge of opposite sign.

---

4. To get a direct feeling for how this works, experiment with two disk or ring magnets of the sort found in toy stores or often placed on refrigerator doors. Note how they orient themselves relative to each other when brought into proximity.

Ionic bonding is most important between atoms that differ greatly in how strongly their nuclei grasp their surrounding electron clouds, a characteristic known as *electronegativity.*

Most atoms are prone to link together into molecules primarily through the influence of covalent bonding, while the bonding between others is mainly ionic bonding. In general, however, a mixture of ionic and covalent bonding acts to bind atoms together. The bond between two hydrogen atoms joined to make the hydrogen molecule H_2 is almost purely covalent; the one bonding sodium to chlorine to make ordinary salt NaCl is nearly purely ionic, and the bonding of two hydrogen atoms to an oxygen atom to create the water molecule H_2O is about 60% covalent.[5]

These—the cooperating covalent and ionic bonding forces—are the two main glues that join most atoms together to form molecules, and in addition to them are two other much weaker but important bonding forces that involve both atoms and molecules. **Figure 2.2B** schematically illustrates the four bonding forces. One of them, called the *hydrogen bond*, owes its power to the special characteristics of the hydrogen atom— it is simply one proton tied to one electron. The electron swarms of all atoms larger than hydrogen (and all other atoms are larger) tend to form electrical shields around their nuclei that reduce their ability to project strong electrostatic forces outside the atoms. Unlike all these other nuclei, the positively charged hydrogen nucleus has virtually no shield, and so it is left hanging out almost naked when the hydrogen atom's electron covalently pairs up with an electron owned by another atom. The electrostatic field of that nearly naked proton reaches out to pull toward itself any negative charge that might be in the vicinity. Because, like all electrostatic coulomb fields, the proton's field falls away with the square of the distance, it is too weak to form a chemical bond with most atoms. However, the field is strong enough that if the naked proton can approach an electron pair owned by another atom, the electrostatic attraction between the proton and those electrons is sufficient to form a weak bond. This, the hydrogen bond, is typically less than one-tenth as strong as the covalent bond, and it mainly forms only with small atoms like oxygen, nitrogen, and fluorine. Hydrogen bonds are of such a strength that their formation and rupturing play rich instrumental roles in the orchestra of life on this planet, and perhaps elsewhere.[6]

---

5. Pauling (1960) 84–90.
6. Pauling (1960) 450–51.

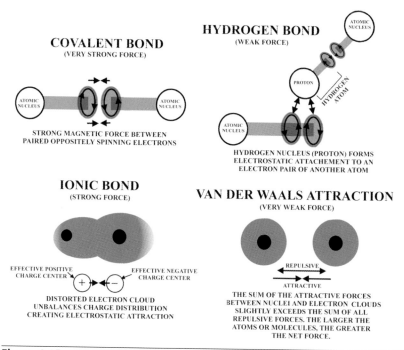

Figure 2.2B Schematic illustration of the four bonding forces important in binding atoms and molecules together.

The Four Kinds of Chemical Bonds Compared			
Covalent Bond (also called electron sharing)	Ionic Bond (electrostatic bond, also called valence bond)	Hydrogen Bond (occasionally called the *vital* bond)	Van der Waals Attraction
Very strong	Strong-very strong	Weak	Very weak
Magnetic in nature	Electrostatic (coulomb force) in nature		
Joins atoms together to form molecules.	Joins atoms together to form molecules.	Joins molecules together to from associated liquids or solids.	Makes atoms or molecules in liquid or solid state attract each other.
Special Feature: Widespread, so nearly universal it is sometimes called the chemical bond. It tightly joins atoms to form rigid molecules.	*Special Feature:* Joins individual atoms like sodium and chlorine into huge interlocking crystalline arrays: (Na - Cl - Na - Cl, etc.).	*Special Feature:* Joins comparatively small molecules (like H_2O) into arrays. Likes to hook to oxygen.	*Special Feature:* Important force holding liquids together: strength increases with size of molecules or atoms involved.

The fourth bonding force is even less strong. Called the **Van der Waals attraction**, it is a weak electrostatic attraction between atoms or molecules. Atoms or molecules in the liquid state come close enough together for the mutual attraction between each atomic or molecular nucleus for another's electron swarm to be slightly greater than the sum of the repulsive electrostatic forces between the positive nuclei and those between the negative electrons. The larger the atoms and molecules, the stronger the Van der Waals attraction, and that causes heavier atoms and molecules to have higher boiling points.[7] It is the Van der Waals attraction that must be overcome in the transition from the liquid to the gaseous state, though if the atoms or molecules participate in hydrogen bonding that too must be overcome.[8]

■ Construction and Properties of the Water Molecule

When two hydrogen atoms join with an oxygen atom to create a water molecule (as depicted in **Figure 2.3**) the bonding is strongly covalent; that is, the bonding between each hydrogen atom and the oxygen atom is primarily due to the magnetic attraction between the hydrogen atom's single electron and one unpaired electron in the outer shell of the oxygen atom. The oxygen atom has two unpaired electrons in the outer shell, and hence the ability to join with two hydrogen atoms. It has six other electrons: two in the innermost shell (the k-shell) that do not interact chemically plus two sets of paired electrons. (The electrons in each pair have virtually identical orbits, but they spin in opposite directions and therefore hang together by their mutual magnetic attraction.)[9] When the covalent bonds develop, the new molecular entity consists of an oxygen nucleus core surrounded by a two-electron swarm in the inert inner k-shell and, outside of that (in the l-shell) four sets of paired electrons, two of which have attached hydrogen nuclei. The two-dimensional drawing in Figure 2.3 fails to do justice to the resulting three-dimensional structure of the water molecule, but **Figure 2.4** comes closer to portraying it. There it is shown that, because of the electro-

---

7. The Van der Waals force also operates on solids; for example, it is the force acting across cleavage planes to hold sheets of mica together.
8. Additional discussion in Appendix A.
9. And the same quantum numbers, except for the spin quantum number. See Appendix A.

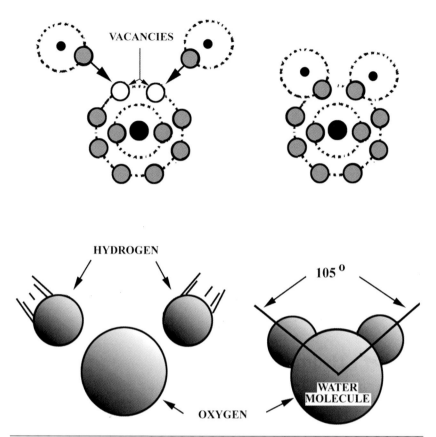

Figure 2.3 The oxygen atom forms covalent bonds with the two hydrogen atoms, creating a very stable molecule not easily torn asunder.

static repulsive force between them, the four sets of electron pairs try to arrange themselves as far apart as possible within the outer shell, a configuration achieved if they are at the four corners of a regular tetrahedron with the oxygen atom at its center. (The regular tetrahedron is a four-sided figure with sides in the form of equilateral triangles. Its corners lie on the surface of a sphere, at points equidistant from each of the other three.) The angles between the lines joining any two sides of a tetrahedron are approximately 109°, but the presence of the two hydrogen nuclei in the structure of the water molecule weakens the repulsion between the bonding pairs of electrons enough to reduce the angle between the lines joining the oxygen and hydrogen nuclei to 105°, as indicated in Figure 2.3.

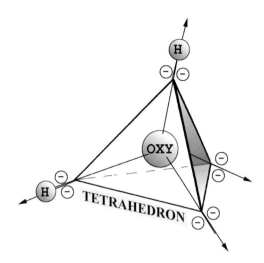

Figure 2.4 In the water molecule two sets of paired electrons and two sets of shared electrons (with their attached hydrogen nuclei) occupy the four corners of a regular tetrahedron with the oxygen atom at its center.

▪ Water's Hydrogen Bonds

The water molecule is one of the few naturally occurring molecules able to participate in hydrogen bonding. This ability results from the relative nakedness of the two hydrogen atoms in the molecule. Though each is tightly held to the central oxygen atom by a strong (magnetic) covalent bond, the two positively charged hydrogen nuclei stick out like little Velcro hooks ready to grasp electrostatically onto any negatively charged objects that come close enough. Favored candidates are electron pairs held in the outer shells of oxygen atoms because these atoms are physically small enough for the electrostatic Velcro hooks to make contact. Other water molecules are a great source of oxygen atoms, and each one has two sets of electron pairs available to be on the receiving end of a hydrogen bond. In Figure 2.4, these are the two pairs at the lower right-hand corners of the tetrahedron, the ones not involved in the covalent bonding that locks the water molecule together.

When water is in the gas state the individual water molecules are too far apart ever to form hydrogen bonds with each other. Thus H_2O truly is the proper formula for steam. But when water is cool enough to condense into the liquid state some of the molecules come close enough together to begin forming hydrogen bonds. Liquid water is not simply H_2O; rather it is $(H_2O)_n$, where $n = 1, 2,$ and 3 or more. Also, n has varying values in different parts of the liquid, and it tends to get larger the more the water cools. As illustrated in **Figure 2.5**, water molecules come together through hydrogen bonding to form temporary and shifting associations, either in pairs, in strings or in rings composed of three to six (or perhaps even more) molecules.[10] Then when water cools enough to freeze into ice, the molecules lock together to form crystal arrays of water molecules held together by hydrogen bonds in a definite fixed pattern. As this happens the 105° angle between the covalent bonds of the water molecule open up to the 109° angle (the angle between the lines drawn from the center to any two corners of the regular tetrahedron) that gives ice a perfectly symmetrical crystal pattern, one called hexagonal and which is said to have cubic symmetry.[11] Diamond has this same pattern, but it is much harder than ice because the attachments between the carbon atoms in diamond are all tight covalent bonds that bring the atoms close together. Ice is much weaker because the bonds between water molecules are all hydrogen bonds, only 4% as strong as covalent bonds, and that gives the ice crystal a very open structure.

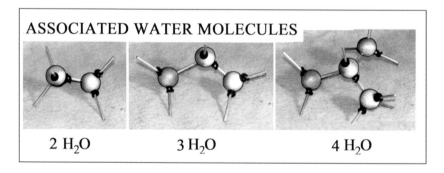

ASSOCIATED WATER MOLECULES

2 H$_2$O   3 H$_2$O   4 H$_2$O

Figure 2.5 Photographs of wooden models of water molecules joined together in temporary liaisons typical of liquid water. Other forms may also exist, such as strings or rings formed of four or five molecules. The wooden balls represent oxygen atoms, the wooden sticks portray the hydrogen bonds, and the black grommets wrapped around the sticks represent the hydrogen atoms in the water molecule.

---

10. Research on the nature of associated water clusters is ongoing. One recent paper [Gregory et al. (1997)] reports recent results on how the clustering affects the dipole moment of water molecules.
11. The cubic symmetry means that it is possible to draw cubes around portions of the crystal structure so that the structure in each cube is exactly like that in every adjoining one.

This tendency of water molecules to clump together (to associate, in technical parlance) through hydrogen bonding when in the liquid state is abnormal; most molecules cannot form hydrogen bonds so they keep the same formula when in the liquid state as they have when in the gaseous state. The strange situation with water is well described by an analogy following after one suggested by the famous British scientist Lord Kelvin. Water molecules can be likened to ships to illustrate their behavior in the solid, liquid and gaseous states. Molecules of water in the gaseous state are like ships on the high seas in that they are far apart and rarely in visual contact. Water molecules in the liquid state are like ships jammed together in a harbor—they are close by each other and some of them are temporarily linked by grappling hooks (the hydrogen bonds) as the ships transfer cargo or await berths. The molecules in the solid state are like ships in drydock, locked into total immobility by hydrogen bonds.

▪ Consequences of the Water Molecule's Shape and Ability To Form Hydrogen Bonds

Water, its strong electrical character

Even though it contains electrical charges, every atom appears to be electrically neutral when examined from a position well outside the atom. That is because the number of positive charges in the atom equals the number of negative charges, and these are symmetrically arranged so that the center position of the positive charges is identical to the center position of the negative charges (both are at the center of the atom). Each charge has its own electric field but in this situation the fields all cancel each other, so the atom looks to have no electrical character; that is, it appears to be electrically neutral. (Of course a person walking inside the atom with an electric field meter would recognize radical changes in electric field with position.)

A molecule might also appear to lack electrical character, but only if, like an atom, it is symmetric so that the centers of its positive and negative charge distributions are colocated. If the molecule is asymmetric so that the charge centers do not coincide, the molecule displays electrical character. Electrically, it looks like a little dumbbell having a positive charge in one end and an equal negative charge in the other. Called an electric dipole because of its two electrical poles (one positive, one negative), this configuration can be described quantitatively by saying that it has a *dipole moment* $m = ql$, where q is the amount of positive or negative charge and l is the separation between the two poles (the charge centers). **Figure 2.6**

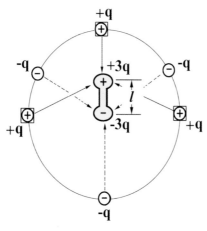

DIPOLE MOMENT = 3q*l*

Figure 2.6 The geometrical centers of the positive charges and the negative charges shown here do not coincide. Electrically, the distribution of charge is equal to that of a dipole having length *l* and charge 3q. Its dipole moment is 3q*l*. Notice that the centers of the positive and negative charge distributions are at the ends of the dipole.

schematically illustrates the relationship between a contrived unsymmetrical distribution of charges and an equivalent dipole.

The water molecule's distribution of charge is unsymmetrical in a fashion somewhat similar to that illustrated in Figure 2.6. As easily seen in Figure 2.4, the location of the center of the water molecule's negative charge distribution differs greatly from the location of the positive center. For that reason, the water molecule has an appreciable dipole moment, but that is only part of the story. If two of the little dumbbell dipoles can be locked together end to end the resulting dipole moment is twice that of the single dipole, and if three can be locked together the resulting dipole moment is tripled. Since liquid water has many hydrogen bonds linking molecules together, it contains molecular assemblies with abnormally high dipole moments.

The importance of water being so abnormally dipolar becomes evident when it is placed in an electric field. As shown in **Figure 2.7**, the molecular dipole arrays each try to swing around to orient themselves within the electric field in a direction that causes their electric fields to oppose the original field. Since electric fields are additive, the end result is a reduction in the overall electric field in the region. The greater the length of the dipoles

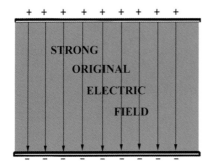 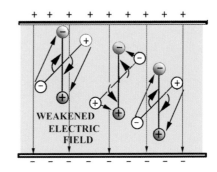

Figure 2.7 The electric field in a vacuum between the plates of a parallel-plate condenser is proportional to the number of (+) and (-) charges on the plates, as at left. When dipoles are introduced, as at right, the electric field pushes the positive end of each dipole toward the negative plate, and the negative end of each dipole toward the positive plate. The longer the dipole arms, the greater the shift in charge, and the more the electric fields of those shifted charges (shaded circles) oppose the original field, thereby reducing the electric field between the plates. The amount by which the field is reduced is defined as the dielectric constant of the introduced material. Water has dielectric constant 80.

introduced, the more profound the reduction to the original electric field, and so water—thanks to the odd shape of its molecule plus its hydrogen bonding—reduces the electric field to a remarkable 1/80 of its former value.

The amount by which a substance reduces an electric field is called its **_dielectric constant_** ε, and so water has dielectric constant $\varepsilon = 80$. The dielectric constant is much easier to deal with conceptually, and much easier to measure than dipole moments, so this is the term normally used in describing this electrical characteristic of a substance—its ability to reduce an electric field in which it is immersed. The dielectric constant of most liquids is in the range 1.5 to 10, so liquid water's dielectric constant of 80 truly departs from the norm. Hydrogen bonding is largely responsible because molecular linkages generally increase the effective dipole moment,[12] and the dielectric constant goes up proportionately.

Water, the universal solvent

Water's abnormally high dielectric constant is the cause of its capability to dissolve so many other substances, so many that it has been called the

---

12. Not quite doubling because the molecular dipoles do not line up exactly end to end when the bonds form. See also Appendix A.

"universal solvent." Because of its high dielectric constant, water virtually destroys the electrostatic bonds, i.e. the electric fields, that hold so many substances together. Only those molecules held together by nearly pure covalent bonding (electromagnetic rather than electrostatic in character) are able to resist water's electrical onslaught. Water's ability to reduce the electrostatic (ionic) bond forces by a factor of 80 causes ionically bonded molecules like salt (NaCl) literally to collapse in water. As salt dissolves in water the molecules break up into positive sodium ions Na^+ and negative chlorine ions Cl^-. Moreover, these ions are attracted to water molecules by electrostatic forces: that between the negative oxygen end of a water molecule and the positive sodium ion, and that between the positive hydrogen end of the molecule and the negative chlorine ion. This additional attraction between water molecules and ions tends to keep the dissolved components of crystals from reforming as they might were the extra electrostatic forces absent. As far as each water molecule is concerned, the attraction is temporary and does not cause any lasting alteration. Thus water molecules remain unchanged and can participate again and again in dissolving other substances. That is important, as is the fact that water is so highly dipolar—otherwise water would not be such a powerful geologic force nor would it be able to carry enough dissolved mineral nutrients to permit vigorous plant growth.

Water, its elevated boiling and melting points and its high heat capacity

A general consequence of hydrogen bonding is the elevated temperature of water's melting point, and also its boiling point, compared to that of similar triatomic molecules such as H_2S and H_2Se (S for sulfur and Se for selenium) that do not normally enter into hydrogen bonding. Without hydrogen bonding, the expected melting point of ice would be −95°C and the boiling point −80°C. If that were to be the case, then the earth's surface now probably would have no water since any around would have been in vapor form and would have escaped into space long ago.

The *heat capacity*[13] of a substance is, by definition, the amount of energy per unit mass required to raise the temperature by 1°C. Water's heat capacity is very high, 1 calorie per gram per degree centigrade. Most elemental and molecular substances have much lower heat capacities, typically

---

13. A related term is *specific heat*, the ratio of a substance's heat capacity to the heat capacity of water. Water has specific heat = 1.

one-half to less than one-tenth that of water. Recall that the temperature of a collection of atoms or molecules is merely a measure of their degree of agitation or motion. For water molecules to take up the increasing motion represented by a given increase in temperature, some of their hydrogen bonds must be broken, and that takes additional energy beyond that required just to increase the motion. This requirement for extra energy is the cause of liquid water's high heat capacity, about twice that of water in solid form (ice) and in vapor form (steam) because raising the temperature of those molecules does not require breaking hydrogen bonds. See **Table 2.1.**

Closely related to liquid water's high heat capacity is the large amount of energy required to break the rigid hydrogen bonds in ice when melting it and the even higher amount of energy required to break all the Van der Waals and hydrogen bonds when converting liquid water to steam. The energy stored in these bonds is called latent heat. The *latent heat of fusion*, 80 calories per gram, is the amount of energy required to break the hydrogen bonds in ice when converting it to water, and the *latent heat of vaporization*, 540 calories per gram, is the amount needed to break all the bonds when converting water to steam.[14] And of course just the opposite happens when steam is cooled down enough to convert it back to liquid water: as the Van der Waals and hydrogen bonds corral the rapidly wandering steam molecules, their energy of motion must be cast off. The liquid molecules still mill around in their confined situation but their motion is now comparatively slow, and it becomes slower yet as cooling continues. Water molecules then suddenly cast off the last of their energy of motion and essentially stop in their tracks as freezing occurs. The amount of energy involved when water changes state is large since, compared to other common substances like acetone, alcohol, sulfuric acid, turpentine and benzene that lack hydrogen bonding, water's latent heats of fusion and vaporization are abnormally high, by factors ranging from 2 to 30. When the ground freezes, water's high latent heat of fusion causes the release of much heat, and if the freezing is to continue that heat must somehow be carried away.

The combination of water's high heat capacity and latent heats together with its plentiful nature makes it the great moderator of climate. Each year, sunlight evaporates about 95,000 cubic miles of water from the earth's surface, and all those molecules racing around in gaseous form

---

14. If ice sublimates directly to vapor, then the heat of sublimation is the sum of the heats of fusion and vaporization.

Table 2.1 Densities, heat capacities and thermal conductivities of various substances, listed within each column in order of decreasing magnitude. (Compiled primarily from data given by Williams and Smith (1989), pages 90 and 109.)

Density (kg/m°)	Thermal Conductivity (Watts/m °K)	Heat Capacity (Joules/kg °K)
Quartz: 2,660	Quartz: 8.8	*Water: 4,180*
	Clay Minerals: 2.92	Unfrozen Peat with
Clay Minerals: 2,650		80% $H_2$0: 3,600
	Ice: 2.24	
		Unfrozen Peat with
Sandy or Clay Soil:	Frozen Soil: 2–3	40% $H_2$0: 3,300
1,500 to 2,000		
	Unfrozen Sandy Soil with	*Ice: 2,100*
	40% $H_2$0: 2.20	
Organic Matter: 1,300		Dry Peat & Dry Organic
	Frozen Peat: ~2	Matter:1,920
Water: 1,000	Unfrozen Sandy Soil with	Unfrozen Clay Soil with
	20% H_2O: 1.80	40% $H_2$0: 1,480
Ice: 917	Unfrozen Clay Soil with	Unfrozen Clay Soil with
	20% H_2O: 1.58	40% H_2O: 1,150
Peat: 300 to 1,100	*Water: 0.56*	Unfrozen Sandy Soil with
		20% H_2O: 1,180
	Unfrozen Peat with	
Air: 1.2	80% H_2O: 0.50	***Air: 1,010***
	Unfrozen Dry Sandy	Clay Minerals: 900
	Soil: 0.30	
		Unfrozen Dry Clay
	Unfrozen Peat with	Soil: 890
	40% H_2O: 0.29	
		Dry Sandy Soil: 800
	Unfrozen Dry Clay	
	Soil: 0.25	Quartz: 800
	Dry Organic	
	Matter: 0.25	
	Snow (220 kg/m^2): 0.11	
	Dry Peat: 0.06	
	Air: 0.025	

transport huge amounts of energy from low to high latitudes, making life possible everywhere—again, thanks to the hydrogen bond.

On a smaller scale we see illustrations of water's high heat capacity and latent heats in everyday life:

- The long-lasting warmth of a hot water bottle placed between the sheets, illustrating water's high heat capacity.
- The coolness of a wet shirt on the skin as the evaporating water molecules suck up energy because of water's high heat of vaporization.
- On cold nights the use by citrus fruit growers of water sprayed over their orchards to release energy that will slow the drop in temperature and possibly protect the fruit from freezing. Similarly illustrating the high heat of fusion is the way that the temperature often falls to freezing on a cold night, then pauses there until most of the moisture has frozen out of the air.

Water, its high viscosity, surface tension, and wetting ability

All liquids tend to resist changes to their form. This resistive property, called *viscosity*, is an internal friction or stickiness that increases with the degree to which the molecules of a liquid cling to each other. That is right down water's alley because its hydrogen bonding causes so many of its molecules to stick together that water has higher viscosity than almost any other liquid. Water's viscosity is high near the boiling point, but it increases as water cools because that allows greater numbers of hydrogen bonds to form between the molecules. The viscosity at 0°C is more than six times the viscosity at 100°C.

Related to water's high viscosity is its high surface tension, the tendency of the surface of a liquid to contract. Like viscosity, *surface tension* depends on the attractive forces between the molecules in the liquid. Water has the highest surface tension of all commonly occurring molecular liquids. Mercury, not a molecular substance, has even higher surface tension. In dynes/cm, a unit often used to express surface tension, mercury's surface tension is 470, water's is 73, while most liquids have surface tensions in the range 10 to 40. Some people find it good sport to place bits of mercury on a glass plate since, because of its high surface tension, the mercury beads up and rolls around over the glass in little balls. By comparison, playful people find that water is no fun at all because it tends to spread out over the glass. It is not just a matter of the difference in surface tension; the different behaviors are due to the fact that mercury molecules have no affinity for glass molecules, but water molecules have a strong affinity. Why? Again, the graspy little hydrogen bond is the culprit. It causes water molecules to reach out for oxygen atoms, and glass contains many since its formula is SiO_2. Water also wets other materials that contain oxygen, such as cotton fiber, rock, clay, and both organic and inorganic soil particles.

This tendency for water to adsorb, that is, to stick to the surface of soil particles and plant materials, is of extreme importance to plant growth— and also to what happens to water when the ground freezes. Water's propensity to stick to itself and to the surface of other materials is the crucial element in its ability to rise up through the soil from below, to climb up to the tops of the tallest trees and to crawl right through frozen ground and even layers of ice. But before going into these matters we need to examine one other important consequence of the hydrogen bond, the strange fact that ice floats on water.

Water, why it floats ice

That ice is less dense than water and therefore floats on it is extremely peculiar because virtually every other substance gets more dense when it changes from the liquid to the solid state. Most every liquid steadily contracts as it cools and when it reaches its freezing point suddenly contracts by about another 10%, so the solidifying portion sinks to the bottom.[15]

The reason why liquids contract as they cool is that the Van der Waals forces pull the molecules increasingly close together. As with other liquids, in water the Van der Waals forces pull cooling molecules closer together, but only until the cooling proceeds down to 4°C. At that temperature the influence of water's hydrogen bonds overpowers the influence of the Van der Waals attraction. Any two water molecules held together by a hydrogen bond stand a definite distance apart—no farther and no closer than the spacing at which an exact balancing occurs among all the forces involved. These are mainly the attraction between the hydrogen nucleus and the oxygen's electron pairs, and the repulsion between the hydrogen nucleus and the oxygen nucleus. This spacing, 2.76×10^{-8} cm between each two oxygen atoms, is more than twice the spacing normally attained between other non-metallic atoms when chemically bonded or crystallized. The carbon-carbon spacing in diamond is only 1.54×10^{-8} cm, the oxygen-oxygen separation in

---

15. Among the few substances that do not contract upon freezing are antimony and bismuth alloys, which makes them useful as casting metals because a slight expansion on freezing creates crisp outlines, as on type face. Pure bismuth expands 3.3% upon solidification.

molecular oxygen is but 1.2×10^{-8} cm, and that between carbon and hydrogen in paraffin is 1.1×10^{-8} cm. (However, the spacing between crystallized metallic atoms may also be large, ranging to more than 5×10^{-8} cm.)

The transition at 4°C is smooth: it is just that at this temperature the standoffishness of the hydrogen bond reverses the contraction of water upon further cooling. Some of the molecules continue to move closer together as the temperature falls, but more and more of them are pushing farther apart as they form new hydrogen bonds. As the cooling progresses, the water molecules continue to mill around like a crowd of soldiers relaxing just prior to a battalion review. Some stand alone while others clump together, but below 4°C they are almost imperceptibly moving apart slightly and gradually becoming more orderly because of the pressure from increasing numbers of hydrogen bonds. Then, after the temperature has fallen a few more degrees, it is as if a command barks out and the molecular soldiers stiffen their hydrogen-bond arms and push apart, freezing into regular ranks. Each then stands the hydrogen bond-length 2.76×10^{-8} cm from four others to create a perfectly symmetrical pattern. A key element of the pattern is the open hexagonal ring-like structure illustrated schematically in **Figure 2.8** and with photographs of a wooden model in **Figure 2.9**. The six oxygen atoms (with their attached hydrogen atoms) in each hexagon do not all lie in the same plane, rather they lie in a corrugated sheet—one having regular ups and downs as indicated in Figure 2.8. Stacked one above the next, these corrugated ring arrays form ice crystals with much open space between the molecules, most easily seen in Figure 2.9. In the upper left part of Figure 2.9 the oxygen atoms marked with an "X" all lie at one level in a corrugated sheet, and those marked "O" lie in the other level.

The view along the direction identified as the c-axis of the crystal (see sketch at lower right in Figure 2.9) shows the most openness, and a view along any of the three a-axes shows almost as much, since these views also are looking through a series of hexagonal rings shaped slightly different from those seen looking down the c-axis. The view taken midway between two a-axes illustrates mainly the tiered nature of the crystal, i.e., the stacking of one corrugated sheet above another. Notice that the views along the three a-axes in the basal plane[16] are identical, and that these are 120° apart. Thus, an ice crystal rotated 120° around its c-axis merely moves every oxygen atom into the position formerly occupied by another.

---

16. The basal plane is perpendicular to the c-axis. Ice crystals tend to grow fastest in the basal plane [Knight (1967)].

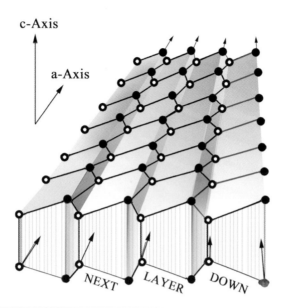

Figure 2.8 The main features of the structure in an ice crystal are hexagonal rings of water molecules lying in corrugated sheets. These lie perpendicular to the *c*-axis of the crystal, and the ring structure also is seen in a view looking along an *a*-axis. Compare with Figure 2.9.

Ice crystals melt and grow preferentially in the *a*- or *c*-axis directions according to the circumstances, so the identification of the axes is more than a matter of just wanting to give names to things. However, for the purpose here the primary significance of the structuring depicted in Figures 2.8 and 2.9 is the open nature of the ice crystal, and that openness is what makes ice 11% lighter than water (its density is 0.917 that of water). Hydrogen bonding is the basis for the openness because it causes each oxygen atom in an ice crystal to have only four nearby neighbors, and these cannot get very close because the hydrogen bonds grasp them and hold them at bay with stiff, outstretched arms. Curiously enough, the amount of energy given off as ice melts indicates that the melting involves the breaking of only about 15% of the hydrogen bonds holding an ice crystal together.[17] This means that water near the melting point is a highly **associated liquid**; that is, a high majority of the water molecules are joined together into hydrogen-bonded multiple arrays.

---

17. Pauling (1960) p 468.

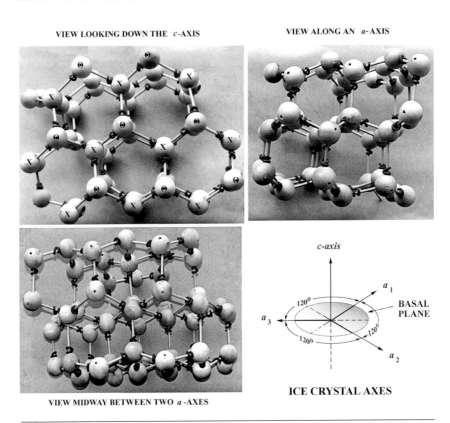

Figure 2.9 Photographs of a wooden model depicting the structure of ordinary ice. Large balls represent oxygen atoms, and adjacent rubber grommets the hydrogen atoms.

■ Supercooling, the Consequence of the Universal System Happiness Rule

Although ice always melts at 0°C, water does not necessarily freeze at that temperature. Perfectly pure and still water can remain liquid when cooled to approximately –40°C. The barking command that initiates the freezing can be a stirring, a shaking, or the introducing of a tiny crystal that acts as a nucleating agent. If no command is given by the time the temperature reaches –40°C, the molecular troops jump to dress-parade positions by their own volition.

Any substance that remains liquid when cooled below its melting point is said to be supercooled. *Supercooling* is a common phenomenon in nature, and in fact it is not amiss to say that every liquid always undergoes some supercooling prior to freezing. A person occasionally hears about supercooling but almost never about what really causes it. Supercooling is too important in the field of permafrost to let go unexplained, and to help with that I introduce what I call here the Universal System Happiness Rule:

▪ Universal System Happiness Rule ▪
Every System Is Happiest When It Contains The Least
Possible Amount Of Free Energy

My name and phraseology for this rule is merely a mnemonic gimmick intended to anchor in the reader's mind this crucial idea that governs all things: the formation and behavior of atoms and molecules, why permafrost is as it is, why an ice cream cone melts before you can get it all in your mouth, and whatever else that you can think of. This rule appears repeatedly in this book, and even if left unstated the rule is still there governing what happens—or fails to happen, such as water not always turning into ice as the temperature falls through the melting point, 0°C.

The Universal System Happiness Rule really is an expression of the Second Law of Thermodynamics, about which one author[18] of a college-level physics text said, "Unfortunately all formal statements of this law are given in technical terms that require lots of explanation...." Though perhaps requiring lengthy explanation, some of the Second Law's familiar consequences are that water runs downhill, heat moves from hot places to cold places, and perpetual motion machines never really work.

The name "Universal System Happiness Rule" conveys the general idea but a more suitable and rigorous statement of the principle is: *The stable, equilibrium state of a system is the state of minimum free energy.* Two terms in this statement require explanation.

System.   The term "system" here is in the sense of Webster's first meaning of the word: "a regularly interacting or independent group of items forming a unified whole," and more specifically "an assemblage of substances that is in or tends to equilibrium." In essence, a system is some portion of matter that stands alone, essentially free of outside influence. Our solar system is an example, as would be an atom floating around all by itself far

---

18. Weniger (1940).

from others. A system can also be some portion of matter in between these two extremes that is sufficiently isolated that we can consider it as an entity unto itself. Examples are the hot cocoa contained in a thermos bottle, a spinning top, or perhaps a pair of ring-shaped magnets constrained to encircle a vertical wooden stick. Another example more directly related to the topic of this book is a bucket containing partially frozen moist soil.

Free energy.   Free energy is available energy; that is, the energy that a system possesses that is available to do work. The amount of free energy in a system depends on the mass of the system, its temperature, and perhaps the location of the matter of which the system is composed. The free energy also depends on the system's heat capacity—that is, the ability of the system to carry heat energy. (Recall that water has very high heat capacity compared to most other substances: a kilogram of water will carry far more heat than a kilogram of soil or rock.)

A system only possesses free energy when it is capable of doing work, and it is only capable of doing work if it will change spontaneously into a configuration of greater stability. A system that is stable—and therefore in possession of minimum free energy and also unable to do work—can be changed into an unstable system if work is performed on it or it receives a transfer of free energy from another system.

Back in 1886 the famous Austrian scientist Ludwig Boltzmann made note that the struggle for life in all its forms is a battle for free energy. The main source of free energy here on earth is sunlight. Some of that energy is used by plants to build up unstable organic compounds that are the source of free energy in food, and some sunlight converts to mechanical energy that powers the motions of the atmosphere and the oceans. Were it not for all that free energy arriving from the sun, the earth would be a miserable place to live. The solar energy is free not in the sense that we do not have to pay for it, but rather free in the sense of not being locked up, and therefore available to do work.

Just why every system adjusts itself (by whatever means is available) to a condition of minimum free energy seems, at first glance, rather mysterious. What magic allows the system to know when it has achieved a state of minimum free energy? Actually, like a magician's trickery, the magic is merely illusory and always has a common-sense explanation. For example, imagine a system consisting of a marble that happens to be placed on an absolutely level table which, at one place, has a bowl set into the table so that its lip is exactly at the level of the tabletop. While the marble is resting on the tabletop it is stable and unable to do any work—it has minimum free energy. Now imagine that the marble is placed on the lip of the bowl in the table. It is still at the same level as before, but now it can do work by rolling down the side of the bowl. Relative to the bottom of the bowl, the marble has free energy in the form of potential energy, the energy of position. Where does the marble finally come to a stop? The answer of course is the bottom of the bowl, the place where it has achieved a minimum of free energy, the place where it is stable, and the place where it no longer can do any work. (You might well ask what work was involved in the marble's rolling down the bowl. It did not actually do any useful work other than to stir up the air in the bowl and perhaps heat it slightly, but had a fly been in its path, the marble could have performed enough work to flatten the fly. The point is that having free energy provides the capability to perform work, whether or not the work actually gets performed.) And of course it is easy to see how the marble apparently knew it had reached a state of minimum free energy when it got to the bottom of the bowl. Obviously it had lost what free energy it had, the potential energy directly proportional to the height of the bowl, and so the marble could do nothing except rest happily in the bottom of the bowl forever.

Another simple example is a system consisting of a weight suspended from a spring. The system achieves a minimum of free energy when the downward force of gravity pulling on the weight just balances the upward force created by the stretching of the spring. This system is stable and it is not capable of doing any work. But now suppose that an external agent (like a person's hand) pulls the weight down some distance. The pulling involved doing work on the system, an amount equal to the force required times the distance the weight is moved. The system now is not happy: it is unstable and it has free energy in the form of potential energy, the energy of position. That free (potential) energy is equal to the work done on the system, and it is available to do the same amount of work. When the weight is released, the system will get rid of its free energy and oscillate back to its former stable, equilibrium state, the state of minimum free energy.

The Universal System Happiness Rule, *The stable, equilibrium state of a system is the state of minimum free energy,* applies to whatever system is at hand. That system might even be a horribly complicated one that involves matter in solid, liquid and gaseous forms and that is influenced by all conceivable kinds of interactions, chemical, mechanical, and electromagnetic.

When liquid water freezes it is merely minimizing its free energy, in keeping with the Universal System Happiness Rule. As water cools, its decreasing thermal motion permits a growing number of its molecules to

clump together through hydrogen bonding (to associate), and those associated molecules possess less free energy than do the remaining unattached molecules. At temperatures above the freezing point, the free
energy carried by unattached molecules and the lesser free energy of the
associated ones add up to a free energy less than that of ice, hence the
water remains liquid. However, when a given mass of water becomes
supercooled, that is, remains unfrozen even though the temperature is
below 0°C, its free energy is actually *greater than* the free energy of an
equal mass of ice. According to the Universal System Happiness Rule, this
is an unstable situation, and it would seem that the water should have
sought stability (happiness) by becoming frozen. Nevertheless, the system
is stable, but just barely, because a certain tiny amount of work must be
done if the water molecules are going to solidify into ice. Going back to
the analogy of the milling soldiers, we might think of the work necessary
to the forming of ranks as akin to the exertion involved in the soldiers'
stiffly extending their arms. This extra work relates to the fact that as soon
as water begins to convert to ice the overall system changes from being
one of only liquid water to one consisting partly of water, partly of ice and
partly of the interface between the two. Physically, the interface consists of
the ice's outermost layer of water molecules, which recent studies are indicating is sort of half-molten and half-solid, with the vibrational motions
of the molecules three to four times greater than those of molecules
deeper in the ice crystal,[19] and free energies correspondingly higher as
well. This interface has relatively high free energy associated with it in an
amount that depends on the area and shape of the interface. If the interface is very small and has high curvature (as would be true if the ice were
in the shape of a tiny sphere) any growth of the interface causes the overall
system free energy to increase, so conversion of water to ice does not
occur. Yet if the temperature drops low enough, or if a foreign object with
radius bigger than what is called the **critical radius** is introduced into the
liquid, freezing can occur without increasing the system's overall free
energy. Then, in fact, freezing does decrease the system's overall free
energy. The mathematics of the situation is given in Appendix A.

Herein lies the explanation of why a tiny foreign particle introduced
into a supercoooled liquid may cause sudden freezing. The foreign particle substitutes for ice in forming a nucleus for further growth and so
nucleation is the term applied to this process. The foreign particle must

19. Seife (1996).

have a radius at least as large as the critical radius (typically about 10^{-7} cm) so that the energy associated with building the ice-water interface around it no longer is a barrier to further freezing.[20] Stirring the liquid may also work because agitation can increase the free energy of the water locally. At that local spot the water then may have enough extra energy to supply that needed to make the ice-water interface grow until it reaches the critical radius. In essence, free energy is borrowed by one part of the liquid from other parts, and even though an increase in free energy occurs in the borrowing part, the overall free energy decreases. Once the critical radius is reached, the whole mass of liquid water may quickly freeze.

When water does freeze, an amount of energy equal to water's latent heat of fusion must be cast off. The latent heat is the difference between the free energy carried by the mixed collection of solo and clumped molecules and that of molecules locked into the ice crystal lattice.

One consequence of that slight expansion of water as it cools below 4°C is that the temperature in the lower reaches of a deep lake can never fall below 4°C. The heaviest water, that at 4°C, sinks to the bottom of the lake, and so the freezing of a lake is from the top down.[21] Even lakes located in permafrost areas do not freeze completely if they are deep enough. The lack of freezing is crucial to the survival of some aquatic plants and animals, and it can have profound influence on the topography of lake shores in permafrost areas.

Another consequence of water's expanding at temperatures below 4°C is that increasing pressure causes water's melting point to decrease rather than to increase, as with most substances. For this reason it has long been suggested that the pressure created by an ice skater's blades provides a layer of water on which to glide along. (However, the so-called premelting phenomenon to be discussed in Chapter 3 contributes to forming a water layer and, except near 0°C, that contribution is more important in aiding the progress of the skater.) Similarly, the pressure from the weight of a glacier on the ice at its bottom is suggested to create a fluid layer there for the glacier to slide upon, and also make the lower part of the glacier plastic so that

---

20. Nucleation—sometimes called seeding—works best if the introduced particle has a crystal structure similar to that of the solid to be formed. Because of similar crystal spacing and shape, silver iodide crystals are effective for seeding clouds to form ice which then can melt as it falls, and perhaps collect more water to form raindrops big enough to reach the ground before evaporating.
21. However, in a fast running river the mixing of cold and warm water may allow ice to form on the riverbed or within the stream.

it can deform internally. The mechanism is thought to be a collapsing of some of the hydrogen bonds in ice so that some molecules return to the liquid state, even if the temperature is below 0°C. These molecules then are supercooled and subject to refreezing once the pressure is relieved, a process sometimes referred to as pressure-induced *regelation*. An often-cited example of the process is the slow movement downward of a weighted wire through a block of ice—the wire cuts through leaving no obvious sign of its passage. Related also to the rupturing of hydrogen bonds in ice's open crystal lattice is a process called *translation* wherein only a partial melting occurs, but it is enough to allow ice crystals to slide (translate) past one another, and thus for the ice to undergo plastic flow. Pressure alone probably does not explain all observed regelation and translation processes since some appear to be driven by temperature gradients.

The Universal System Happiness Rule governs these and all other things. Examples, some from matters previously discussed, include:

- An electron and a proton will come together to form a hydrogen atom because by doing so they will have less free energy than they had standing alone. Similarly, two hydrogen atoms will join to form the stable hydrogen molecule H_2 because that molecule has less free energy than two hydrogen atoms standing alone. The water molecule forms because it has less free energy than one oxygen atom and two hydrogen atoms standing alone. Similarly, the oxygen atom's outer shell has two sets of paired electrons and two unpaired ones because that arrangement gives the atom lower free energy than it would have if all those six electrons were paired.
- The 105° separation between the water molecule's projecting hydrogen atoms occurs because that configuration has less free energy than any other involving one oxygen atom and two hydrogen atoms. Likewise, the spacing between the oxygen and hydrogen atoms gives the molecule less free energy than any other spacing.
- Water molecules clump together by hydrogen bonding (they associate) because by doing so they have less free energy than when standing alone.
- And, somewhat in reiteration: Water turns to steam because, above the boiling temperature, water molecules in the gaseous state have less free energy than they would have if still liquid. Similarly, water freezes into ice whenever the total free energy of the ice-water system (including the energy associated with the interface) is lowered

by the freezing. If the freezing requires an increase in overall system energy the water will supercool even though the free energy of ice is less than the free energy of water at the temperature of the system. When the interface between ice and water has high curvature (small radius) the free energy associated with the interface is very high, so it takes a lot of energy to make the interface grow. Once the interface reaches the critical radius its further growth always reduces the overall free energy of the ice-water system.

- Dogs lie down to sleep because they have less free energy in that position than when standing up. That's why sleeping dogs are happy dogs, and it's best to let them lie.

▪ Chapter Summary

When atoms join together to form molecules, two main bonding forces hold them together. One, called covalent bonding, depends on the powerful magnetic force that acts between two nearby but oppositely spinning electrons, each belonging to separate atoms. Another strong bonding force that causes atoms to join to form molecules is electrostatic in nature: the negative charges of one atom attract the positive charges of the other to bind the atoms together. Most molecules actually are held together by a combination of these two strong bonding forces, but when the hydrogen atom is involved yet another force can operate. Called the hydrogen bond, this relatively weak electrostatic attachment binds the hydrogen component of a molecule like water to certain other molecules or atoms that are small enough to allow the hydrogen nucleus to approach and electrostatically attach to one of their electron pairs. Oxygen is such an atom, so it is a likely partner in hydrogen bonding, and this fact has important consequences.

In part because of its ability to form hydrogen bonds, water has abnormally strong electrical character that makes it the universal solvent. Water's hydrogen bonding not only enhances its electrical nature, it causes water to have abnormally high viscosity, abnormally high surface tension, and proclivity to cling strongly to itself and to many other substances, especially those containing oxygen in their makeup: glass, dirt, and organic matter. These characteristics are crucial to the explanation of what happens when soil freezes.

Unlike almost all other substances—thanks to hydrogen bonding—water expands as it freezes. It also typically undergoes supercooling, the phenomenon of remaining liquid below the melting temperature, 0°C. The explanation of how that comes about involves the important concept of free energy and the fact that systems are most stable when containing minimum free energy. Called here the Universal System Happiness Rule, this principle comes up again repeatedly in this book as we now move ahead to delve into the behavior of the ground when it freezes and thaws.

When the Ground Freezes

▪ Introduction

A person need not live where there is permafrost, but only where the temperature falls below freezing at least a few nights each year, to observe first-hand some of the intriguing and important phenomena that can make permafrost such a treacherous material. The key here is the movement of water molecules through the ground during and after freezing. When water freezes, its volume expands by 9%, but when water-saturated soil freezes, its expansion might exceed 50%, 100%, or even more. Such radical change could not happen if the soil were isolated (say, if placed in a steel or plastic bucket). However, extreme expansion can and does take place in natural settings where fine-grained soil is in contact with underlying material that provides a water supply. The water migrates upward into the freezing soil and turns to ice, thereby increasing the volume of the soil, the phenomenon known as frost heave.

▪ Pipkrakes, the Crunchy Product of Ground Freezing

A vivid and easily seen example of the consequences of water's ability to migrate to where the soil is freezing is the formation of **pipkrakes**. At one time or another most everyone has walked out on a frosty morn and experienced the pleasure of padding through the pipkrakes. These little pillars of ice crunch nicely beneath the feet. Also called **needle ice** or mushfrost, as well as Kammeise in German, and shimobashira in Japanese, pipkrakes are vertically elongated crystals of ice that grow just beneath the ground surface. One cold night (–2°C or lower) might produce a layer of piprakes one or more centimeters thick. Several cold nights in succession interlaced by daytime warming will produce multiple layers of pipkrakes, creating an array 10 cm or more in thickness. A prolonged spell of temperature below –2°C can cause the individual needle-like pipkrake crystals to grow to lengths exceeding 40 cm (15 in).[1] **Figure 3.1A** contains examples found at sea level in Washington state, an area better known for its apples than its frozen ground, but pipkrakes grow there, too. **Figures 3.1B** and **3.1C** show Alaska pipkrakes.

Figure 3.1A Pipkrakes formed during several cold nights at sea level on San Juan Island, Washington. The arrows mark the lower edge of the pipkrake layers. The two photos at bottom show pipkrakes removed from the location shown in the photograph at upper right. See also Plate 3.

1. Schmid (1955), cited by Washburn (1980); Tricart (1970).

BASE

Figure 3.1B Pipkrakes photographed by Richard Reger in Alaska. Notice the needle-like character of those in the upper photo and the more massive character of those below. Arrows mark the nightly layers formed.

Figure 3.1C   Is this the grandaddy of all pipkrakes? Billy Conner, a researcher with the Alaska Department of Public Facilities and Transportation is shown with a substantial specimen. Perhaps formed by another mechanism, this large ice mass protruding approximately 1 meter above its surroundings was photographed by Richard Veazey near the trans-Alaska pipeline on Alaska's North Slope. See also Plate 4.

Pipkrakes grow from the bottom up, initially starting out one to several centimeters below the ground surface. The growth then lifts that surface layer upward. Three primary factors intertwine to determine the growth of pipkrake crystals: the temperature regime, the soil characteristics, and the water supply. For the crystals to continue their growth the temperature of the air in contact with the ground must be low enough to carry away the energy released by the forming ice (the latent heat of fusion, 80 cal/gm). Plenty of water must be available in the soil below where the pipkrakes are forming, and the soil there needs to have characteristics that promote the movement of the water up to the base of the pipkrake layer. Minor differences in the amount of organic matter on the soil surface can have major effects by altering the thermal conductivity of the material, so pipkrake fields may contain many irregular pillars of ice that lift some parts of the ground surface well above others. The ground then becomes karst-like on a microscale.

Pipkrakes tend to grow in loose soil, and they tend to loosen the soil, so their presence is a self-perpetuating phenomenon that can have major effect on the soil surface. One consequence in areas of high wind is that the formation of pipkrakes, by lifting up and loosening fine-grained soil, can lead to the phenomenon known as soil *deflation*, the removal of the smaller grains (clay, silt, and sand) by wind action. The effect is sufficiently deleterious on some agricultural lands that farmers sometimes try to compact pipkrake-prone soils.

Of general geomorphological significance is that pipkrake formation can contribute to *cryoturbation*, a churning of soil that can cause sorting by particle size and differential motions, especially on sloping ground. The pipkrake needles grow perpendicular to the ground surface, so if the surface slopes, the growth and subsequent thaw of the needles will tend to transport any lifted material downslope. Also, the needles often develop curvature under the load of surface material carried, so even on level ground material lifted by the needles may fall back to a new position. On sloping ground, pipkrakes sometimes cause displacements of several to several tens of centimeters during the course of a year.[2]

Pipkrakes can help sort near-surface material in two ways. As seen in **Figure 3.2**, the formation of pipkrakes tends to lift the small-sized soil material up away from pebbles or rocks. Also, if rocks lie atop the ground the formation of pipkrakes can lift the rocks above the surrounding surface. Canadian geocryologists have reported the lifting of stones weighing as much as 10 kg.[3] Measurements of downslope movement caused by pipkrake formation show a variety of results, but in most instances the measurements indicate that the needle ice growth moves fine-grained material much faster than coarser material and therefore helps to sort it. Geocryologists think this action contributes to the formation of a variety of *patterned ground* features, among them stripes which extend up and down gradual alpine slopes and *cryoturbation steps* which form staircase-like on steeper slopes (25° to 38°) in relatively dry midlatitude areas that experience freeze and thaw.[4] The cryoturbation steps shown in **Figures 3.3A** and **3.3B** are in northeastern Oregon. Such steps are easy to observe in many

---

2. MacKay and Mathews (1974a;1974b); Lewkowicz (1988); Washburn (1980).
3. MacKay and Mathews (1974a).
4. Sharp (1942); Washburn (1980); Lewkowicz (1988).

Figure 3.2 Soil uplifted by pipkrakes away from pebbles in a wet driveway on San Juan Island, Washington. See also Plate 5.

Figure 3.3A Cyroturbation steps (terracettes) on a 35-degree north slope overlooking the Weatherby Rest Stop at Mile 335 on Interstate 84 in northeastern Oregon. The estimated spacing of these parallel steps is 0.5 to 1 m. Photographed in summer during the late 1990s.

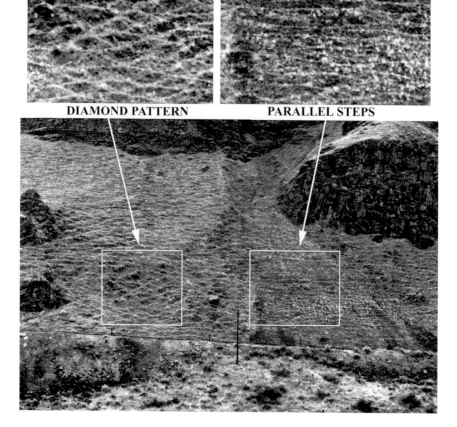

DIAMOND PATTERN   **PARALLEL STEPS**

Figure 3.3B At left, slightly sloping cryoturbation steps crossing over each other to form a diamond pattern, and, at right, a pattern of parallel horizontal steps. The parallel pattern is far more common than the diamond pattern. Limited observations suggest that, when seen, the diamond pattern typically appears near the bottoms of slopes that exhibit the horizontal steps higher up. Photographed March 1999 looking south from Interstate 84 in northeastern Oregon.

lightly vegetated hilly regions of southern Canada and the northern United States.[5] Generally the steps lie parallel to the slope but some slopes display a diamond pattern of inclined steps crossing over each other, as in Figure 3.3B.

5. History buffs take note that on June 25, 1876, General George Custer led his troops through intermittent arrays of cryoturbation steps as he approached the location of his last stand on the hill above the Little Bighorn River in southern Montana. Surrounded by more pressing matters, it is doubtful if the doomed general or any of his men noticed the steps. The upper slopes of Custer's hill are too gentle to support the development of the steps, and the same is true of the nearby hill where Major Marcus Reno's command managed to hold off the Sioux Indians. However, between those hilltops stands the third important prominence of the battle, Weir Point, with one upper slope steep enough to display a few poorly developed cryoturbation steps. Other more well-developed steps appear on the steepest slopes in the battlefield and also north of it on the bluffs facing eastward onto the Little Bighorn.

Pipkrakes grow best in loose soil containing a substantial proportion (greater than 20%) fine-grained particles, but not too many clayey fines because they can inhibit the movement of water up through the soil. This movement of water is crucial to the growth of the pipkrake crystals, and also to the growth of other forms of annual and perennial ground ice found in cold regions. So an important factor in what happens when the ground freezes is the nature of the soil, especially the size of the soil particles involved.

▪ Soil, the Mineral and Organic Matrix

The word *soil*, like so many others, has multiple meanings. In this book I use "soil" in a broad sense to mean the uppermost layer of material at the earth's surface that is perhaps frozen or wherein freeze and thaw processes are taking place. In a narrow sense, "soil" is only the topmost part of this layer, the part that supports plant life. The nature of a soil at any location depends much on climate, its original source material and length of time it has been exposed to weathering processes. A typical soil is a complex of irregularly fragmented particles, both crystalline and amorphous, which are arranged in intricate geometric patterns of uncounted variety. Near-surface soil is always in a perpetual state of change; it swells and shrinks as it wets and dries or freezes and thaws, and it interacts mechanically, chemically, and electrically and with whatever materials fill its pores, be they gases, liquids, or transient solids.

With regard to what happens when the ground freezes and thaws, the single most important characteristic of a soil is the size of its particles. Soil particle size fixes both the size of the pores in the soil and the surface area at the interfaces between the soil particles and the liquids and gases in the pores. If the soil particles are large, the pore spaces are large and the interfacial surface area is small, whereas if the particles in the soil are small, the pore spaces are small and the surface area is large.

Soil texture is the term used to describe the size range of particles in soil. Traditionally, soil scientists divide particles into size ranges called separates or fractions. Going from the smallest to the largest grain sizes, the fraction diameters are: clay, < 0.002 mm; silt, 0.002 to 0.05 mm; sand, 0.05 to 1 mm; very coarse sand, 1 mm to 2 mm; and gravel, 2 mm to 5 cm.[6] Upwards of that are pebbles < cobbles < boulders, with no sharp

---

6. Several different schemes exist, but they all are very similar [Hillel (1980) p 57].

Table 3.1 Soil Texture by Grain Size[7]

Texture Name	Diameter Limits
Boulders > Cobbles > Pebbles	Larger than 2 cm
Gravel	2 cm – 5 mm
Very Coarse Sand	1 mm – 2 mm
Sand	0.05 mm – 1 mm
Silt	0.002 mm – 0.05 mm
Clay	Less than 0.002 mm

distinction between them. See **Table 3.1**. Soils are often described by their predominant fraction; i.e. sand or silt. If no one texture size predominates, the soil is called *loam* if it contains more than 35% clay and less than 27% silt. If the proportion of silt to clay is reversed, the soil is called *silt loam*. Loess, a soil type common in interior Alaska and other periglacial areas, is wind-transported silt and sand (plus minor clay) material mostly in the size range 0.005 to 0.5 mm, with particle size usually decreasing with distance from the original source location. Brown, tan, or gray, loess's color reflects its depositional and postdepositional conditions and processes. Gray loess contains more organic material or ferrous iron than tan loess.

The size distribution and the shapes of the particles in a body of soil are important in determining the soil's behavior, in part because these characteristics affect the packing, i.e., the geometrical relationships between particles. The packing within a box containing marbles of uniform size is different than that if the marbles are of various sizes, and certainly marbles, coins, and wooden matchsticks each pack differently. More important yet is the fact that particle shape and size distribution affect surface area, and the surface area of the particles in a soil profoundly influences how strongly the soil will adsorb water or other substances. Spherical soil particles have the lowest *specific surface* (surface area per unit weight or per unit volume), and platelike particles have the highest. Sand, with its generally spherical particles, typically has specific surface less than 1 m^2/gm, whereas the specific surface of clay may exceed 100 m^2/gm because of the smaller particle size and the platy shape.[8]

---

7. Based primarily on a table by Wimer (1965).
8. Montmorillonite clay has specific surface 800 m^2/gm because its platelets are extremely thin [Hillel (1980) pp 65–67].

Yet other factors come into play to affect the cohesiveness of a soil. A soil might contain only crystalline particles that are completely unattached to each other so that the soil is very loose—for example, a dry desert sand—or it might contain materials that tend to cement the crystalline particles together. Amorphous substances such as humus matter or iron and aluminum oxides and carbonates can act as glues to make the soil cohesive. Shape and history come in here as well, as is evident to a stone mason. He much prefers flat rocks to round rocks when building walls, and even with the flat rocks he knows that flat rocks dumped on the ground result merely in a pile of flat rocks, but if stacked carefully they will form a strong structure, especially if set with cement in the joints.

Adding further to the intricacy of processes that take place in a soil and affect its characteristics is the electrical nature of soil particles, particularly those with irregular shapes. Clay particles, thin flat crystalline entities, typically carry an irregular distribution of charge, and additionally, an excess of negative charge. Both the irregular distribution and the excess of negative charge cause the clay particles to interact electrically with other particles and any substances within a soil's pore spaces. An explanation for the typically negative excess charge on a clay particle is the occasional appearance in the platy crystal structure of a triply charged aluminum ion (Al^{3+}) where a quadruply charged silicon ion (Si^{4+}) normally should reside, and sometimes doubly charged magnesium (Mg^{2+}) ions substitute for the triply charged aluminum ions in the crystal structure. Each substitution creates a dearth of positive charge, with the overall result that the crystal lattice has too many negative charges. Also, unbalanced charges on atoms sometimes appear at the edges of the flat crystal structures, there creating collections of excess positive charge.

Collections of unbalanced charge do not persist in nature because their electrostatic fields quickly attract balancing charges of the opposite sign. Both air and water contain adequate supplies of ions, so the negatively charged flat surfaces of clay particles quickly build a balancing layer of positive ions just outside the surfaces.[9] If a clay particle is in contact with air, the balancing positive layer is thin and tight against the clay sur-

---

9. Air at the earth's surface contains hundreds to thousands of ions per cubic centimeter, and even pure water contains both positive and negative ions because a tiny fraction (one in ten million) of the molecules are disassociated.

face. If the clay is immersed in water, the balancing layer is more diffuse and extends out into the liquid in the pore spaces in a distribution that minimizes free energy.[10] Like clay particles, humus particles in a soil body are typically negatively charged, and therefore they too attract balancing positive charge layers. As in the case of the water molecule, such a particle then has a dipolar charge distribution, because the centers of positive and negative charge are not colocated. Humus particles also act as cementing agents.[11]

Included in the materials attracted to clay and humus particles are the positive ends of water molecules' dipoles. This electrical attraction, along with hydrogen bonding of the water molecules to the soil, explains why clay and humus matter adsorb water so strongly.[12] Another consequence of the electrical nature of clay and humus particles is the development of electrostatic attractive forces between the particles that help draw the particles together. Acting in opposition are electrostatic repulsive forces between soil particle surfaces of like sign. These attractive and repulsive forces and also Van der Waals forces operate with different intensities and vary differently with distance; the way they balance out will help determine if soil particles tend to coagulate or disperse.[13] Affected are both the degree to which soil particles coalesce to form larger aggregate assemblies and the strength of a soil.

In sum, soil is a highly complicated medium composed of mineral and organic materials of various sizes and shapes. The sizes and shapes involved greatly affect the strength of a soil and how it interacts with liquids and gases within the pore spaces. These ongoing interactions are mechanical, chemical, and electrical, and so soil is a complex substance prone to change with time as the soil's environment changes. Typically, the character of a soil depends strongly on climate, and it varies with depth beneath the surface. Additional discussion of these variations is in Appendix B.

---

10. The configuration—negative charge at or just inside the clay surface, and a balancing layer of positive charge just outside—is called an electrostatic double layer. Ions in the double layer can be exchanged for others in a surrounding solution. This ion exchange process is crucial to plant growth because it is a mechanism for supplying the food plants need, the exchange taking place between root hairs and the surrounding liquids or directly between root hairs and soil particles [Hillel (1980) pp 77–83].

11. Hillel (1980) p 77.

12. Hillel (1980) p 85.

13. Hillel (1980) pp 87–90.

■ The Movement of Water through the Ground under the Influence of Suction

The Universal System Happiness Rule's insistence that every system strive to hold minimum free energy provides the fundamental explanation for the formation of pipkrakes. Water transported to the bottom of the pipkrake layer freezes into pipkrake ice because the water molecules possess less free energy when locked into ice crystals than when in liquid or gaseous form. The water molecules move horizontally or upward through the ground toward the freezing front because by doing so they decrease their free energy.

Differences in free energy make things happen: the differences may cause molecules to combine chemically with others, to change state (solid, liquid, or gas), or to move from one place to another. Here we are concerned with change of state of water molecules and the motions they undergo through soils. For this purpose it is useful to think of free energy simply as potential energy. When water falls on the surface of the ground in the form of rain or snow, some of the molecules evaporate back into the atmosphere, some sink into the soil, and others run off along the surface. In each instance, differences in free energy determine the fate of the molecules (even those evaporating because they are supplied with additional free energy from elsewhere).

The simplest fate to describe and understand is that of the water molecules that run off along the surface on the road to lowered free energy. They are moving downhill under the influence of gravity, and are impeded in that motion mainly by the water molecules' high viscosity, large surface tension, and proclivity to cling to the upper surface of the ground—all products of water's ability to form hydrogen bonds.

More complicated is the motion of water molecules that enter the ground. These molecules feel the pull of gravity, but other important forces come into play, and they are very powerful because of the abnormally graspy nature of water molecules, both for each other and for the soil particles they contact. Forces involving **adsorption, capillary** action, and **osmosis** literally suck water down into the ground and later may lift it up again to the surface or higher, into plants and trees. In dealing with movement of fluids and vapors though substances, scientists use various ways to describe quantitatively how these forces operate, and one of the easiest involves the concept of suction.

Before delving into that, it is worthwhile to note that in soils and plants the most important movements of water molecules involve those in liquid form. Water vapors, i.e., collections of H_2O molecules in gaseous form, play a role, especially in the movement of water through unsaturated coarse-grained soils, but the primary transport through fine-grained soils and through plants involves water in liquid form.[14] Hereinafter, unless the context suggests otherwise, the word "water" is used to imply the liquid state of this substance.

▪ Suction and Several of Its Causes

Since all healthy mammals are born with an innate ability to apply suction it is at the outset a familiar idea. Furthermore, suction is measurable and therefore has useful quantitative meaning. The ordinary everyday kind of suction is what gets soda from a straw and gasoline out of an automobile's tank. It operates to lift water when the handle of a shallow-well pump descends. This kind of suction has a very definite limit because the operating force is atmospheric air pressure. The weight of the atmosphere pressing against the surfaces of the soda, the gasoline, and the well water causes these liquids to rise in the tube when suction is applied. The limit to the suction obtained in this fashion is the weight of the atmosphere, at sea level 1.033 kg/cm^2 (10.33 tonnes/m^2)[15] or 14.7 lbs/in^2. The same weight comes from a column of water 10.3 m (33.9 ft) high, and a column of the much heavier liquid mercury 76 cm (29.9 in) high.

If as shown in **Figure 3.4** we place a tube in an open vessel of water—like the straw in the soda bottle—and suck on the tube we are withdrawing air molecules from it and thereby are reducing the pressure inside the tube. The weight of the atmosphere pressing on the open surface outside the tube pushes water up into the tube, and if we create a perfect vacuum in the tube by sucking out all the air the water column rises to a height of 10.3 m = 33.9 ft. The pressure at the top of the column is zero, and at the bottom the pressure is equal to the weight of the atmosphere.

A convenient measure of pressure for many purposes is the ***atmosphere (Atm)***, and it is meaningful as well since everyone can readily

---

14. I stress this point here because it is possible to find literature dealing with permafrost and related topics that places too much emphasis on the role of water vapor and also may misidentify the relevant transport processes. For example, a valuable engineering guidebook by McFadden and Bennett (1991) contains statements about the role of water vapor and osmosis on pages 89–91, 167, and 185–86 that are at variance with other references cited here.

15. 1 tonne = 1000 kg = 1 metric ton.

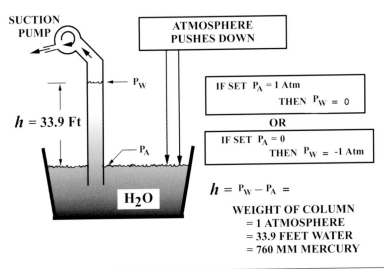

Figure 3.4 When all the air is sucked out of a tube immersed in water, the pressure of the atmosphere will push water 10.3 m (33.9 feet) up the tube.

comprehend its magnitude—imagine, for example, balancing on your hand a stick or other object that weighs 14.7 lb and has an area of 1 in². That's how a pressure of 1 Atm feels. Within the earth, and in the same fashion, the weight of overlying material causes increasing pressure downward; the pressure is near 10,000 Atm at a depth of 30 km, and at the center of the earth it is near 3 million Atm. (In dealing with pressure, scientists may use other units such as the bar, the pascal and newtons/m²; 1 atmosphere equals 1.013 bar, and 1 bar equals 100 kilopascal or 10^5 newtons/m².)

It is obvious that the pressure at the level of the base of the column shown in Figure 3.4 is everywhere 1 Atm, that the pressure declines with distance above the surface, and exactly at the top of the column it is zero. Describing the situation in another way, we can say that a pressure gradient exists within the column, and it is such that the difference in pressure from top to bottom of the column is 1 Atm. That pressure difference (the gradient times the distance) does not depend on what we choose as a reference pressure. For many purposes it is convenient to set the reference level at atmospheric pressure; that is, to choose sea level pressure as zero. Ordinary pressure gauges used to measure air pressure in automobile tires read this way, and if they did not, a person using one would have to mentally subtract 1.033 kg/cm² or 14.7 lbs/in² from the gauge readings, a bothersome chore. Adopting the same idea, and again referring to

Figure 3.4, we can set the pressure at the exposed liquid surface equal to zero. Then Pa = 0, and Pw = −1. The difference in pressure is still the same: Pa − Pw = 0 − (−1) = 1, but we have forced ourselves to take on the concept of *negative pressure*. Suction creates negative pressure.

Now suppose that the suction applied to the tube were only enough to raise the column to height h = $\frac{1}{4}$h$_o$. The pressure at the top of the column would then be −$\frac{1}{4}$ Atm, and we could say that we have applied a suction S = $\frac{1}{4}$ Atm to the column. Doubling the suction to $\frac{1}{2}$ Atm doubles the column length to h = $\frac{1}{2}$. Doubling it again brings us back to where we started h = h$_o$, and suction S = 1 Atm, but in the process we have reaffirmed that suction is a force that creates reduced pressure, and we have seen how it can have precise quantitative meaning. Notice also that since 1 Atm = 76 cm mercury = 29.9 in mercury = 33.9 ft H$_2$0, any of these other lengths also can be used to indicate the magnitude of suction. Vacuum-suction (the suction produced, as Webster says in defining suction, by "the act or process of exerting a force on a solid, liquid or gaseous body by reason of reduced air pressure over part of its surface") cannot exceed 1 Atm = 33.9 ft H$_2$0, but the concept of suction can be extended to describe pressure reductions created by other, more powerful processes such as capillary action, osmosis, and temperature differentials. These processes can create larger suctions (negative pressures) amounting to tens or even hundreds of atmospheres. The important consequence of suctions, regardless of their cause, is that they frequently cause the movement of material from one place to another.

A powerful creator of suction is capillary action, and it is a very important mover of water through soil and plant material. The usual way to illustrate the way capillary action works is to place small-diameter tubes in containers of water and mercury, as shown in **Figure 3.5**. The water rises up the tubes, but the mercury moves *down* the tubes. This difference in behavior is because water wets glass, but mercury does not. With its graspy hydrogen bonds water reaches out to the oxygen atoms in the capillary tube and so not only clings to the walls, it tries to crawl up them so that the surface of the water becomes dished in what is called a concave *meniscus*. Mercury atoms have high attraction for each other (mercury's extremely high surface tension expresses this) but no attraction for glass. Therefore mercury retreats from the walls of a glass capillary tube, sinking in on itself to form a convex meniscus. In establishing these shapes and positions within the capillaries both liquids are merely behaving according to the Universal System Happiness Rule that requires them to arrange themselves to have minimum free energy. Mercury's free energy is minimized by the

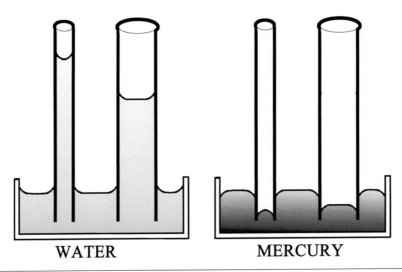

WATER   MERCURY

Figure 3.5 Water rises and mercury falls in a capillary, the distances increasing as the capillary becomes smaller.

element's becoming as spherical as possible because of its high surface tension and its complete lack of attraction for glass, but the free energy of water is minimum when the hydrogen bonds (like Velcro hooks) are enmeshed with oxygen atoms in the capillary walls, and the more hydrogen bonds formed with the oxygen in the walls the better.

In a capillary made of a material that water wets, the water will reach up the wall of a capillary as far as it can, but only to a height that depends on the water's surface tension and on the diameter of the capillary. In crawling up the wall the water must lift its entire mass against the force of gravity using a force that is equal to the water's surface tension σ times the circumference of the capillary $2\pi r$, where r is the radius of the capillary. The opposing downward force is the weight of the water in the column and that is equal to the volume of water multiplied by its density ρ and the acceleration of gravity g. Since the volume is the height of the column h times the area πr^2, the downward force is $\pi r^2 h \rho g$. Balancing the upward and downward forces ($2\pi r \sigma = \pi r^2 h \rho g$) requires that the height of the column be $h = 2\sigma/r\rho g$, and so we see that the height of the column is directly proportional to the surface tension and inversely proportional to the radius.

The pressure in the water just at the meniscus (P_{liquid}) is reduced below the pressure in the air just above (P_{air}) by an amount equal to the downward force per unit area generated by the column, and that is $h\rho g = 2\sigma/r$. Thus $P_{air} - P_{liquid} = 2\sigma/r$ is the suction produced by the capillary action. Notice that as the radius r of the capillary is made smaller both the suction and the height of water in the capillary increase. In principle, both would approach infinity as the radius of the capillary approaches zero. Even if other limiting factors did not interfere, the water column should eventually break despite its great tensile strength. Remarkable as it seems, water in thin columns does have great tensile strength, as great as that of some steel. Laboratory tests indicate a practical limit near 130 Atm, more than twice the tensile strength water needs to pull itself all the way up through the capillaries of even the tallest trees.

Water's ability to move so far up glass capillaries or others such as those in plants that contain ample oxygen in the capillary walls is made possible by the clinging of the water to the wall surfaces, the process called adsorption, and which in the case of water is fostered by its proclivity for hydrogen bonding. (Adsorption is very different from absorption, the process of one substance taking another into its internal structure.) Van der Waals forces play a key role in adsorption, and water adsorbs well on certain substances for yet another reason: in addition to making hydrogen bonds, the water molecule is highly dipolar (that is, it has a large dipole moment, as is indicated by water's large dielectric constant), and that causes water to adsorb to materials like clay and humus particles that also are dipolar. It is an electrostatic attraction between the poles of opposite sign in the adjacent dipolar materials. Silica gel is another strongly dipolar material, and that is why packets of silica gel are included in shipping containers surrounding items that need to be kept dry.

Crucial to the amount of water adsorbed onto another substance is the available surface area. Since the ratio of surface area to volume increases the more finely a material is divided, the smaller its particles the more it adsorbs. A single rock weighing one pound has a surface area far less than 1 ft^2, but if that rock is ground up into fine clay its surface area increases to several tens of acres. Finely divided charcoal, much used to purify air of odors and poisons, has a surface area exceeding 100 acres per pound (90 hectares/kg).

Adsorption can be onto a smooth surface but it is especially effective on pitted and rough surfaces like those of soil particles because their nooks and crannies increase surface area and also take up the role of capillaries. In fact, it is thought that adsorption typically depends much on capillary action, but the fundamental cause of adsorption is the collection of all the intermolecular forces that are in operation at the interface between the adsorber and the adsorbate. When molecules of water or other adsorbed substances stick to surfaces most tightly they are, in essence, experiencing a suction created by hydrogen bonding, the electrical attraction between dipoles of the involved substances, Van der Waals attractions involving the molecules of the water and the adsorber, and electrostatic forces arising from distributions of surface charge. This suction is typically even larger than the purely capillary suction. (Additional discussion of how water's high viscosity, high surface [interfacial] tension, and wetting ability affect its behavior in capillaries and at interfaces appears in Appendix A.)

Since the tightness with which molecular forces at interfaces and capillary action hold water to a soil is of major importance to plant life, soil scientists have devised ways to measure how easily a soil gives up adsorbed and capillary water. One direct way is place the soil on a fine porous plate in a container that can be pressurized to force water from the soil into the atmosphere, as illustrated in **Figure 3.6**. The pressure required to force water out is equal to the suction force holding the water in, so at any pressure the gauge reading directly indicates the suction holding whatever water remains in the soil. A plot of the data obtained as the pressure is

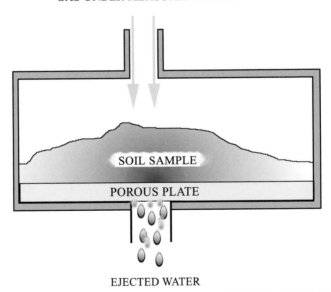

Figure 3.6 Apparatus for measuring the soil moisture characteristic curve of a soil. The gas pressure directly measures the suction.

increased gives for a soil sample a curve of water content versus suction. Several examples of these *soil moisture characteristic curves* appear in **Figure 3.7**, which shows that suctions of 10 or more atmospheres are required to remove the first half of the water contained in various clay soils. The most loosely held water, that held in the soil by capillary action in the larger pores, comes out first, and an increasing proportion of that remaining is the tightly held adsorbed water. In general, the soil moisture characteristic curves show no obvious sharp transition between capillaries and oxygen-rich or dipolar adsorptive surfaces, a signal suggesting that the underlying physical causes of capillary action and adsorption are one and the same: forces acting between molecules lying near each other at boundaries between different substances.

Plants also have another tool for collecting water, osmosis. Osmosis is the one-way movement (diffusion) of a substance through a membrane that is tight enough to exclude some molecules but porous enough to allow the passage of others. Plant and animal cell walls are examples of such semipermeable membranes. An easy experiment displaying the effect of osmosis and how

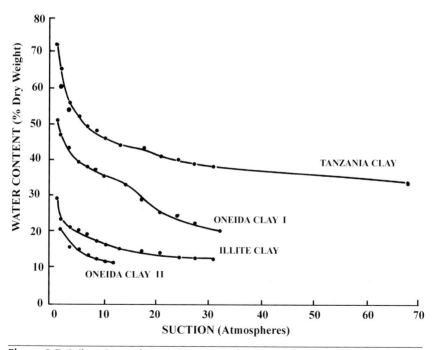

Figure 3.7 Soil moisture characteristic curves for several clay samples. Redrawn from Figure 7.6 of Williams and Smith (1989).

to measure the suction it generates is shown in **Figure 3.8**. A membrane made of skin, parchment, or cellophane separates sugar water from pure water. Sugar does not pass through the membrane, but water does, so the solution rises up the tube to some distance h. The distance h, after correction for capillary effects, is a measure of the osmotic suction. It is found to be directly proportional to the concentration of sugar above the membrane. Familiar examples of osmosis are the passage of nutrients through the walls of the stomach and intestine to the bloodstream, and both the inward diffusion of oxygen and the outward diffusion of carbon dioxide through the linings of the lungs. Osmotic suctions in plants can be very large; the largest observed are in saltbush leaves, in excess of 200 Atm.[16] Clay soils may also in some instances behave in osmotic fashion, causing minor and usually insignificant differential movements of solutions of differing character.[17]

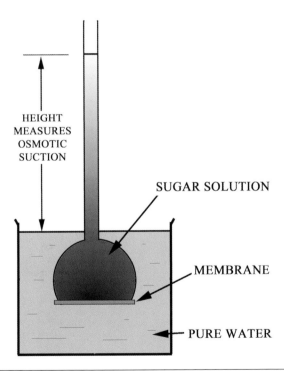

HEIGHT
MEASURES
OSMOTIC
SUCTION

SUGAR SOLUTION

MEMBRANE

PURE WATER

Figure 3.8 Water passes through the membrane, but sugar does not.

16. Meyer (1973). Even higher suctions (lower negative pressures, down to -330 Atm) have been observed in laboratory situations.
17. Hillel (1980) pp 245–46.

▪ Effects of Adsorption, Capillary Suction, and Osmotic Suction in Plants and Soils

In situations such as depicted in Figures 3.4, 3.6, and 3.8, a movement of liquid or gas takes place until equilibrium occurs. It is a transient movement, lasting only long enough to bring the overall system to minimum free energy by transporting a sufficient number of molecules to locations of least free energy. However, if other processes act to remove molecules from the low free-energy end of the system, and to supply them to the other end, flow will continue. In fact, a higher extraction than supply rate generally will increase the suction within the system, until either the supply matches the extraction or the supply limits the transfer through the system.

Such effects are thought to explain how water can rise up through even the tallest of all trees, the giant redwoods exceeding 110 m (360 ft). *Transpiration* from leaves rapidly removes water so that osmosis suctions in the range 5 to 40 Atm typically occur. These large suctions act in conjunction with those produced by adsorption on the walls of plant tissues and capillary action in plant structures. The resulting huge pressure deficits (suctions up to 50 Atm) essentially suck water up through the tree; lifting water to such heights is possible because water clings to itself so well (has such high surface tension), behaving like thin steel wires in small capillary passages.

As plants lose water to the atmosphere by transpiration most of them try to replenish it by sucking water in through their roots, and that effort is strongly affected by soil *capillarity* and other processes that dictate the rates at which water moves through the soil. Water falling on dry soil rapidly diffuses downward until the water molecules adhere to all the surfaces of the soil particles and partially fill the intervening pore spaces. The pore space is about 30% in sandy soils, and about 50% in clay soils. After several days of rain, the rate of water movement downward slows, and the water content becomes fairly uniform in the wetted soil layer. This soil typically also contains much air in the pore spaces unless the soil lies below the water table, and in that case the pore spaces fill with water and the soil is said to be saturated. Most plants grow best in partially aerated soil that is able to release water to the plant roots, but a certain portion of the water adheres so tightly to the soil particles that it is unavailable. Adsorption and perhaps some capillary action locks up about the first 20% (by volume) in clay soils but only the first 5% in sandy soils. One consequence is that sandy soils easily and quickly make water available for plant growth, whereas clay soils release water more slowly. In their pore spaces clay soils are capable of containing more water than sandy soils, so even though they hold back a greater proportion, they may supply more in the long run.[18]

In saturated soils the free energy is fairly uniform throughout, hence pressure deficits between locations are generally less than 1 Atm. The deficit may be near 15 Atm in soils that have released all the water they can to plants (all but the first 5% in sandy soil and 20% in clay soil), and if the water content falls below those levels the pressure deficits (the suctions) get very large, in the range of hundreds of atmospheres. The free energy of the remaining water molecules is so low that plants are unable to suck them away from the soil particles. Even with such large pressure deficits, water usually cannot move through soil fast enough to satisfy the needs of plant growth. When the water cannot come to the plant, if it is to avoid wilting the plant has to go to the water, and that is what happens. It sounds unbelievable, but reported measurements on rye grass indicate that the average rye plant increases the length of its root system by about 5 km (3 mi) each day, and by doing so gains enough water to replace that lost through transpiration.[19]

To recapitulate: although osmotic suction is relatively minor, capillary suction and the suction produced by adsorption are important forces affecting the motion of water through soil. Another powerful suction, *cryosuction*, also comes into play when the temperature drops below freezing. Like the other kinds of suction already discussed, cryosuction can be very large, ranging to up above 200 Atm, so it too will move water through the ground, permitting the formation of pipkrakes and other kinds of ice segregations.

As we move forward it is useful to keep in mind that since suction—however it is caused—is a reduction in pressure from some base level, it can be called a pressure deficit, or a negative pressure. Both terms appear in the literature on soil science, along with others like "pore water pressure," "potential," "matric potential," and "Gibbs free energy." The plethora of terms used to describe what is really only one thing is confusing. The key thing to remember is the direct inverse relation between the free energy of a substance and the suction it experiences. A quick look at **Figure 3.9** may be most helpful of all.

---

18. Hillel (1980) pp 164–65.
19. Meyer (1973).

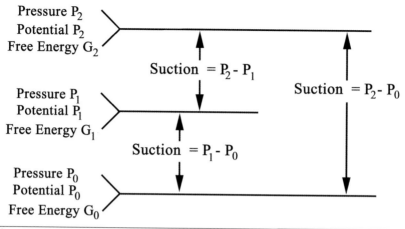

Figure 3.9 A diagram showing the relationship between suction and its related quantities, pressure, potential and free energy.

▪ Cryosuction in Freezing Soil and the Formation of Segregation Ice

Introductory comments

Soil temperatures. Changes in air temperature produce changes in ground temperature, but the ground responds relatively slowly. The situation is the same as that when a person holds one end of a steel bar in a fire.[20] The end of the bar in the fire starts getting hot right away, but it takes some time for the heat energy going into the end of the bar to travel down its length and burn the holder's hand. The temperature within the bar varies with position along the length of the bar, and thus a temperature gradient exists within the bar. And so it is with soil as the temperature in the air changes; the top layer of soil responds the quickest, and the deeper layers follow suit, but more slowly. If a step change in temperature is imposed at the top of the ground, the change will propagate down at a rate that varies with the square root of the time elapsed. Thus if change induced at the top of the ground can penetrate to a depth of 10 cm in one day, the change will penetrate to 20 cm in 4 days, to 50 cm in 25 days, and to 1 meter in 100 days. That is why cold-area residents sometimes are

---

20. John S. Wettlaufer, in a private communication with the author in 1999, suggested using this analogy.

chagrined to have their sewer lines freeze up in early summer, long after the coldest winter weather.

Water migration. Because of the slowness of the ground at depth to respond to changes in air temperature, soil temperature in cold climates is rarely uniform throughout. To put it another way, the soil almost always has a temperature gradient. Seeking to minimize its free energy, any water in the soil will try to migrate through the soil toward the coldest part—that is, down the temperature gradient. Visual proof that water molecules in vapor or liquid form seek the cold is the condensation—or even ice—that forms on the inside surfaces of single-pane windows on cold days. The cold window panes act as sinks that can literally suck water out of the air because the water molecules lower their free energy by collecting on the cold window surface. Much the same thing takes place in a freezing soil in that water molecules travel to and into the soil and collect there in the form of ice.

Freezing point depression. The normal freezing point of water, T_0, is the temperature at which **bulk water** freezes, i.e., water not in capillaries or otherwise confined. $T_0 = 0°C$ when the water is at a pressure of 1 Atm. Increasing the pressure on most substances causes their freezing points to increase, but water is an exception. Its freezing point gets lower with increasing pressure, but only by a slight 0.0074°C per Atm. Consequently the freezing temperature of any bulk water lying several meters below the ground surface is only a tiny fraction of a degree lower than that at the ground surface, although if it were buried beneath several thousand meters of overburden the freezing point would be depressed by 1 to 2°C. Salts or other impurities added to water also lower its freezing temperature. Adding 2% (by weight) sodium chloride lowers water's freezing point about 1°C, and a 22% salt solution has freezing temperature just below –20°C. Water in soil contains some salt but so little that its freezing temperature is not significantly lower than 0°C. Thus in dealing with the freezing of the ground near the earth's surface, depressions of the freezing point T_0 caused by water's salt content or pressures from overburden are of little consequence, and we can think of T_0 as essentially equal to 0°C. Soil begins to freeze when the temperature falls only slightly below 0°C and, as further cooling occurs, more and more water freezes out.

The freezing of soil

If an open container of water is cooled while being stirred so that it everywhere maintains the same temperature, then each gram of water gives up

1 calorie per degree of cooling (its heat capacity) until it reaches the freezing point, 0°C, where the sudden change of state to ice releases 80 cal/gm, the latent heat of fusion. But if that same water is contained within the pore spaces of a soil and adhered to the surfaces of the soil particles, the freezing behavior is quite different. Observations show that the water begins to turn to ice at 0°C, but that some unfrozen water remains in the soil down to quite low temperatures.

Plots of the unfrozen water content of a soil versus temperature such as those shown in **Figure 3.10** are called *soil freezing characteristic curves*. As the water freezes, it gives off a known amount of heat (the latent heat of fusion); thus, by measuring the heat given off from a soil sample as its temperature is changed, it is possible to construct its soil freezing characteristic curve. Notice that the general shapes of the soil

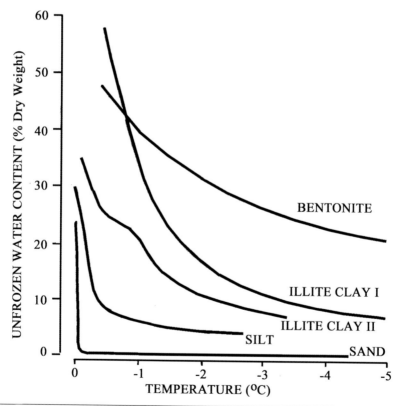

Figure 3.10 Soil freezing characteristic curves for a few clays, a silt and a sand. Redrawn after Figure 1.4 of Williams and Smith (1989).

freezing characteristic curves in Figure 3.10 are very similar to the shapes of the soil moisture characteristic curves shown in Figure 3.7. The similarity is not accidental; in fact, the shape of a soil's freezing characteristic curve can be predicted from its soil moisture characteristic curve. Both curves depend directly on the size of the pores in a soil and the total surface area available for adsorption of water. An extreme example of a soil that resists having its unfrozen water driven off by imposed pressure or lowered temperature is bentonite. This fine clay, formed from weathered volcanic ash, has such small pores and high surface area per unit weight that it retains 20% (by weight) unfrozen water even at −5°C.

By comparing a soil's moisture and freezing characteristic curves, it is possible to determine how much water is retained in a freezing soil at a particular temperature and hence to make an experimental plot relating cryosuction to temperature of the freezing soil, as in **Figures 3.11A and B**. (Another way is to measure the water content directly by one of several different methods, including one known as pulsed magnetic nuclear resonance.)[21] The results are rather stunning because it is evident that cryosuction changes in direct proportion to the temperature as it falls below 0°C, and that the rate of change in cryosuction is very high, approximately 11 to 12 Atm/°C. Thus if a parcel of fine-grained soil has a temperature of −1°C, the cryosuction within that parcel is approximately 11 to 12 Atm, and if another parcel has temperature −10°C, the suction there is 110 to 120 Atm.

The task now is to explain why the variation of cryosuction with temperature is so great, why water and ice can coexist in frozen soil, and why frozen soil is found to contain layers of pure ice called *segregation ice* that typically increase in thickness with depth, as shown in **Figure 3.12**. Toward this end, geocryologists have taken several theoretical approaches, all really based on evaluation of the free energy of a soil-water system to see how it varies when changes are made and certain conditions imposed.

Useful at the outset—and relatively easy to come by—is a general insight into why such high cryosuctions develop in a freezing soil. The starting point is the Universal System Happiness Rule—*the stable, equilibrium state of a system is the state of minimum free energy*. To that rule we apply the principle of Le Chatelier which states that if a system is in a state of stable equilibrium and if one of the conditions is changed then the equilibrium will shift in such a way as to tend to restore the original condition. By itself, Le Chatelier's principle does not help much, but thanks to

---

21. Williams and Smith (1989) p 189.

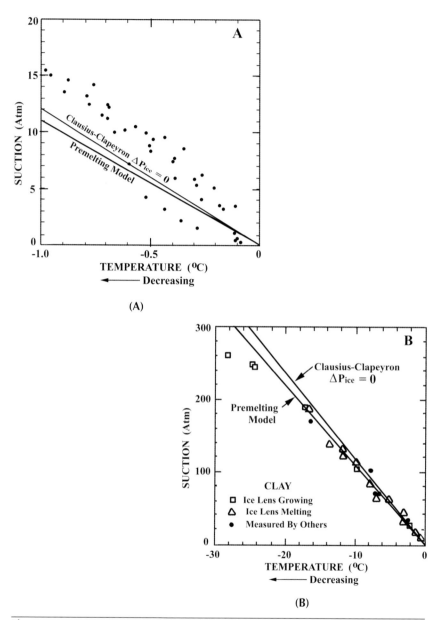

Figure 3.11 (A) Measured suctions in the temperature range −1°C to 0°C and the suctions predicted by the Clausius-Clapeyron equation (with P_{ice} held fixed) and premelting theory. The data points represent measurements on various fine-grained soils, as given in Figure 7.7 of Williams and Smith (1989). (B) Measured suctions in the temperature range −30°C to 0°C together with the suctions predicted by the Clausius-Clapeyron equation and premelting theory. Based on Figure 9 of Dash et al. (1995).

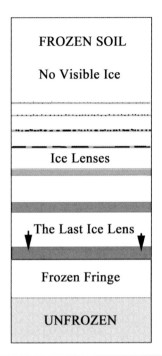

Figure 3.12 Segregation ice often forms in repeating layers that tend to thicken with depth. Based on a diagram (Figure 8.8) reproduced by Williams and Smith (1989).

the work of two gentlemen named Clausius and Clapeyron, we have the Clausius-Clapeyron equation that gives the principle specificity. If a system is in stable equilibrium, and if a change is made to one of the conditions that determines the system's free energy—temperature (T), volume (V) or pressure (P)—the Clausius-Clapeyron equation describes how the other conditions must vary as the system tries to restore itself to stable equilibrium. Applying this equation (derived in Appendix A) to the situation depicted in Figures 3.10A and B—the coexistence of ice and water in a soil of subzero temperature—yields the relationship:

$$\Delta T = (V_{water}\Delta P_{water} - V_{ice}\Delta P_{ice})\, T_o/L. \tag{1}$$

where (T is change in absolute temperature T, T_o is the absolute temperature at the melting point of ice (273 K = 0°C), V_{water} and V_{ice} are the specific volumes of water and ice, ΔP_{water} and ΔP_{ice} are changes in pressure on the water and ice, and L is the latent heat of fusion of water. Suppose that we apply Equation 1 to a system where the pressure on the water P_{water} varies with temperature while the pressure on the ice P_{ice} is fixed by giving it a set value, perhaps that on the soil. Then $\Delta P_{ice} = 0$, and by rearranging some of the terms, Equation 1 becomes

$$\Delta P_{water} = L\,\Delta T/T_o\,V_{water}.$$

Then setting $\Delta P_{water} = P_o - P_{water}$ and $\Delta T = T_o - T$, where P_o is the pressure on the water at $T = T_o$, the normal freezing point of bulk water, $0°C = 273$ K, yields an expression describing how the suction on the water varies with temperature:

$$\text{Suction} = P_o - P_{water} = L(T_o - T)/T_o V_{water}. \tag{2}$$

Since V_{water} is the specific volume of water it equals unity, so the suction depends only of the heat of fusion and the temperature. To see the effect of this expression we can let $T = -1°C = 272$ K, then the suction $P_o - P_{water} = L(1)/273$. The heat of fusion of water is 334 million Joules per cubic meter (or 80 calories per gram), so the suction amounts to 12.2 bars which equals approximately 12 Atm. Thus the difference in pressure on the water and the ice changes by approximately 12 Atm per degree change in temperature, a direct consequence of the combination of the Universal System Happiness Rule and the principle of Le Chatelier. The lowered pressure on the water creates a suction that can draw water into the freezing soil, but again notice that the suction does not depend on the characteristics of the soil; instead it depends only on the heat of fusion of water and the temperature. The straight-line curve labeled "Clausius-Clapeyron" on Figures 3.11A and B represents Equation 2.

If instead of being perfectly constrained, the pressure on the ice is allowed to vary along with the pressure on the water, the pressure difference $P_{ice} - P_{water}$ also varies somewhat. For example, if both P_{ice} and P_{water} drift away from the initial pressure P_o by equal amounts, then $P_{ice} - P_{water}$ = 11.4 Atm/°C, and if P_{ice} changes five times as much as P_{water} when the temperature falls below 0°C, then $P_{ice} - P_{water}$ = 11.1 Atm/°C. The more nearly the pressure on the water P_{water} is constrained the more nearly the pressure difference $P_{ice} - P_{water}$ approaches the limiting value 11.0 Atm/°C. Thus, according to the Clausius-Clapeyron equation, $P_{ice} - P_{water}$ must always have a value ranging between 11 Atm/°C and 12 Atm/°C, regardless of how the pressures P_{ice} and P_{water} drift farther apart with decreasing temperature below 0°C.

The capillary model of segregation ice formation

Segregation ice typically forms in distinct layers of microscopic to multi-meter thickness that can more than double the volume of a soil, causing it to push even heavy objects upward, the phenomenon known as frost heave.[22] The formation of segregation ice also is responsible for the creation of certain landforms and ground patterns found in areas where freezing occurs. One of the first approaches to the explanation of how segregation ice forms and why it is observed in layers within frozen soil is the capillary (or curvature) model of segregation ice formation.

The main idea behind the capillary model is that the relationship between cryosuction and water content in a frozen soil reflects the distribution of pore sizes in the soil. As the temperature of a water-saturated soil falls below 0°C the first water to convert to ice is that least attached to the soil by capillary action and adsorption to the soil particles. It is the water with the highest free energy; that is, it is the water experiencing the least suction, the water in the largest of the pore spaces. If the temperature continues to fall, ice invades successively smaller pore spaces as the free energy of the water molecules contained there exceeds the free energy they would have in the form of ice. The capillary model allows prediction of the size of the soil pores that can be invaded by ice at a given temperature.

Recall from the previous section the relationship derived to find the suction produced by capillary action, $P_{air} - P_{liquid} = 2\sigma/r$, where σ is the liquid's surface tension (interfacial tension $\sigma_{water-air}$) and r is the radius of the capillary. In the capillary model of segregation ice formation the assumption is made that ice-water interfaces behave the same as water-air interfaces, and that if a soil contains ice the pressure on the ice P_{ice} is also the pressure on the soil P_{soil}. In an unburied soil the pressure on the soil is just that of the atmosphere, generally taken as equal to zero. In the capillary model, the suction is the difference between the pressures on the ice P_{ice} (or P_{soil}) and the water P_{water}, the appropriate *interfacial tension* is that between ice and water, $\sigma_{ice-water}$, and the relevant radius is that of capillaries at the ice-water interface $r_{ice-water}$. Hence

$$\text{Suction} = P_{ice} - P_{water} = 2\sigma_{ice-water}/r_{ice-water} \qquad (3)$$

22. The increase in volume in a parcel of freezing soil wherein segregation ice is forming is due to the increase in number of molecules (soil particles plus water molecules), not the expansion by 9% of water as it turns to ice. Although engineers ignored his work for many years, as early as 1918 Taber (1929) performed experiments showing that frost heave created substantial volume changes and that the amount of heave depended on soil grain size.

A tenet of the capillary model also is that the suction $P_{ice} - P_{water}$ is given by the Clausius-Clapeyron equation when the pressure on the ice is held fixed, that is, by Equation 2. Combining Equations 2 and 3 yields an expression for the ice-water radius

$$r_{ice-water} = 2\sigma_{ice-water}T_o V_w/L(T_o - T) \qquad (4)$$

which specifies that the radius of curvature at the ice-water interface must decrease with decreasing temperature below the freezing point of bulk water, a radius assumed to be determined by pore size. Thus ice invades smaller pore spaces as the temperature falls, but the water in those pores with radii smaller than the radius $r_{ice-water}$ remains unfrozen. **Figure 3.3**[13] illustrates the relationships among pore size, suction and temperature, according to the capillary model.[23]

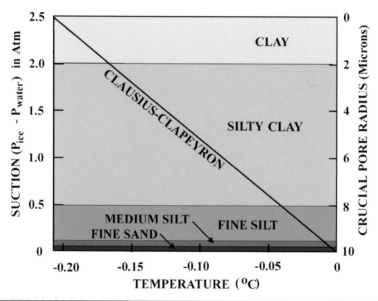

Figure 3.13 A diagram showing, according to the Clausius-Clapeyron equation, how suction varies with temperature, and the minimum size of the soil pores that ice can invade (right-hand scale), according to the capillary model of segregation ice formation. Also shown is the relationship between pore size and soil texture. In soil, the suction increases with declining temperature below 0°C. At a temperature of –0.1°C, the suction is about 1.2 atmospheres and the crucial pore radius is approximately 5 microns, within the domain of silty clay. Based on data given in Tables 6.1 and 6.3 of Smith (1985a).

23. Williams (1999).

As the temperature drops below 0°C and the soil begins to freeze, the capillary model calls for the water in the largest pore spaces to freeze first. At a given temperature, if the pore radius $r_{ice\text{-}water}$ is smaller than the minimum pore size then ice can enter the pores and so the freezing front advances into the soil. However, if $r_{ice\text{-}water}$ is larger than the minimum pore size, the freezing front stalls. In that case, the water at the freezing front is at lower pressure than the water in the soil below, and therefore that water moves toward the freezing front in order to lower its free energy, the situation depicted in **Figure 3.14**. The movement of the water to the freezing front brings to it water of higher pressure then P_{water} in Equation 2 above, and so that water with its higher free energy converts to

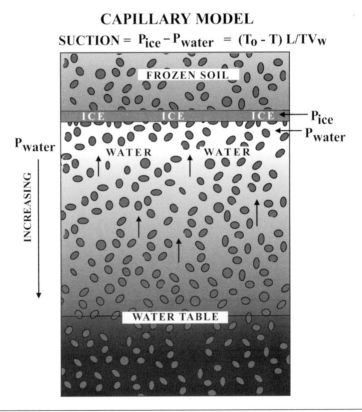

Figure 3.14 A schematic diagram illustrating the capillary model of segregation ice formation. According to this model, when the freezing front stalls, segregation ice forms, the suction (pressure difference between that of the ice and the adjacent water) increases, and water moves upward toward the freezing front. Pressure here is represented by shading; the higher the pressure, the darker the shading.

ice, forming a layer of segregation ice. The heat of fusion given off by the freezing water fights against further drop in temperature, helping to stall the freezing front.

An explanation of why segregation ice forms in multiple, regularly spaced layers, as shown in Figure 3.12, is that the movement of water from the unfrozen soil to the freezing front desiccates the unfrozen soil, thereby lowering the gradient in suction and slowing the movement of water to the freezing front. Less latent heat of fusion is given off, and that may allow the temperature at the forming ice lens to become lower, thereby permitting the freezing front to advance until conditions again become ripe for the formation of another layer of segregation ice. As the freezing front moves down through the soil to greater depth, the changes in temperature become slower, allowing more time for layers to thicken.

The formation of segregation ice in a soil causes its volume to increase, and that volume increase causes the surface of the soil to heave upwards. If the soil in which the ice is forming is essentially at the surface of the ground, then only the pressure of the atmosphere resists the heaving, the situation usually described by setting $P_{ice} = 0$. The pressure on the water P_{water} is the same when the temperature is 0°C, but it becomes more negative as the temperature falls, or alternately, we could say that the suction increases. As long as P_{water} is negative, the suction draws water to the freezing front, segregation ice forms, and the ground surface heaves upward. However, if the freezing soil is buried deep enough, or by other means a heavy enough load is placed on it, the pressure on the soil and the ice P_{ice} increases. The pressure on the water P_{water} must do likewise, always having a value determined by P_{ice} and the temperature. If P_{ice} is high enough at a given temperature to bring P_{water} up to zero, no longer is there a difference in pressure (a suction) between the water at the freezing front and the water in the unfrozen soil. Thus, the movement of water to the freezing front ceases, shutting off further frost heaving. Examination of Figure 3.13 shows that if the capillary model fully describes the formation of segregation ice and its consequent frost heaving, then ice formation could not take place at P_{ice} greater than about 2.5 Atm. This ***shut-off pressure*** occurs when the temperature at the freezing front is approximately –0.2°C.

A soil layer approximately 3 m thick produces an overburden pressure sufficient to shut off the capillary model's formation of segregation ice.[24, 25]

---

24. Smith (1985a).
25. Although individual mineral soil particles can have densities up to 2.7, bulk soil usually has density only 1.0 to 1.7 times that of water.

However, field observations have shown that segregation ice can form in soil buried more than 30 m, and one observation has been made of the ice forming at a depth exceeding 100 m.[26] Also, laboratory experiments show the growth of segregation ice at suctions far greater than allowed by the capillary model (see Figure 3.11B).

The hydrodynamic model of segregation ice formation

The inability of the capillary model to explain the formation of segregation ice at temperatures below about –0.2°C or heaving pressures greater than approximately 2.5 Atm led to development of the hydrodynamic model. Another consideration was that observations were showing that segregation ice was growing at locations behind the freezing front, whereas the capillary model called for formation of the ice only at the front. If segregation ice is to form behind the freezing front it is obvious that water must be moving across the freezing front through the partially frozen region shown as the *"frozen fringe"* on Figure 3.12. That flow of water presumably is through the unfrozen pores and through unfrozen water lying at soil particle surfaces or perhaps boundaries between ice crystals, a deduction forced by observations that frozen soil and ice layers do transmit water when suction is applied.

The approach in the hydrodynamic model is to examine the flow of heat and mass in a freezing soil, recognizing that these are interrelated, and assuming that the only mass transport is that of water. The procedure is to develop an expression for the transport of heat through the soil and another to describe the flow of water, then to link these together by means of the Clausius-Clapeyron equation and the assumption that the temperature dictates the amount of unfrozen water in the soil, that is, by the soil's freezing characteristic curve. The result is that segregation ice should form at some location behind the freezing front where the effective *hydraulic conductivity* changes most rapidly with temperature. At that point, water flowing through the frozen fringe piles up and turns to ice. At temperatures above 0°C the hydraulic conductivity of a soil varies slowly with temperature, but **Figure 3.15**, a somewhat schematic plot of hydraulic conductivity versus temperature, shows that the hydraulic conductivity of a fine-grained soil changes rapidly as the temperature falls a few tenths of

---

26. Wernecke (1932) reported segregation ice veins in quartzite rock at 100–130 m depth at Keno Hill, Yukon Territory. The overburden pressure was approximately 24 Atm, and even higher pressure may have been required to fracture the quartzite.

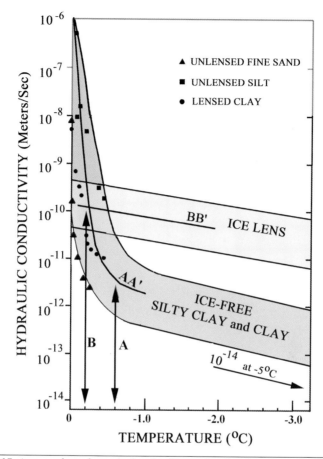

Figure 3.15 A somewhat schematic drawing illustrating that the effective hydraulic conductivity of silty clay and clay is higher than that of an ice layer at temperatures just below 0°C, but at temperatures below about −0.3°C water passes more freely through the ice layer than the silty clay and clay. The hydraulic conductivity changes drastically at the temperatures marked A and B, as described further in the text. Data points are taken from Figure 7.10 of Williams and Smith (1989).

a degree below 0°C, and then declines more slowly at lower temperature. On Figure 3.15 the dark band shows the range in hydraulic conductivity of ice-free silty clay and clay, and the line AA′ represents a hypothetical average silty soil. The data points are the results of measurements made on a fine sand and also on a clay-silt, both ice-free and containing ice. The rapid change in hydraulic conductivity with decreasing temperature occurs to the left of the double-arrowed line marking the temperature

where the slope of the line AA' changes drastically, and so the hydrodynamic model calls for the formation of segregation ice in this hypothetical
soil somewhere in the temperature range 0°C to –0.6°C. For actual soils
the range is typically –0.2°C to –0.3°C.[27] The band marked "ice lens" on
Figure 3.15 represents the range of effective hydraulic conductivities of ice
lenses, its width reflecting the estimated range, and the line BB' a hypothetical average. At temperatures near 0°C the effective hydraulic conductivity of fine-grained soils exceeds that of ice, but at lower temperature the
reverse is true.[28] The hydraulic conductivity of the silt is more than 10
times that of pure ice at –0.1°C, but then as the temperature goes lower,
the effective hydraulic conductivity of the silt falls dramatically until at
temperatures below –0.5°C it is below that of ice by a factor of 10 or
more.[29] Hence if ice is present, water likely will pile up where the temperature is approximately –0.2°C, at least if the line marked BB' on Figure 3.15
represents the hydraulic conductivity of ice reasonably well.

Whereas the hydrodynamic model can explain the development of
segregation ice behind the freezing front, it still fails to account for the formation of the ice at temperatures more than a few tenths of a degree
below 0°C, and also the high heaving pressures that have been observed.
At best, the model allows for heaving pressures of not more than a few
atmospheres.

The rigid-ice model of segregation ice formation

Like the hydrodynamic model, the rigid-ice model seeks to explain the formation of segregation ice behind the freezing front by examining the coupled flow of heat and mass. Ice transport is envisioned to take place by the
process of thermally induced regelation. Driven by a temperature gradient,
the process is as depicted in **Figure 3.16**. In this model the ice forms a rigid
lattice that is separated from the soil particles by a thin layer of adsorbed
water. In the thermal regelation process water molecules move from warmer
to colder locations within the system by continuously transforming between
liquid and solid phases. Water continuously transforms to ice at the
upstream ice boundaries where the free energy of the water slightly exceeds

---

27. Smith (1985a).
28. Loch and Kay (1978); Horiguchi and Miller (1983) both cited by Williams and Smith (1989) p 215.
29. Oliphant et al. (1983); Ohrai and Yamamoto (1985); Smith (1985) all cited by Williams and Smith (1989) p 213.

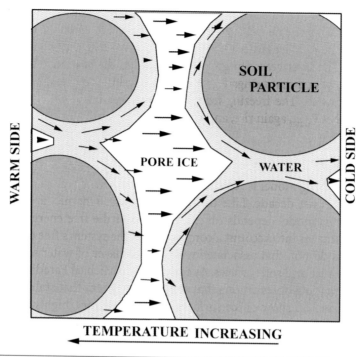

Figure 3.16 According to the rigid-ice model, the flow of water through soil is enhanced at temperatures just below 0°C by thermal regelation, the process involving melting of ice on a warm boundary and refreezing on the opposite cold boundary.

that of the ice at the temperature there, and the reverse occurs at the downstream ends where the ice melts to water. At the warm boundary of an ice body, water traveling in liquid form through the pore spaces transforms to ice, and at the cold boundary the ice transforms back to water. The Clausius-Clapeyron equation is assumed to assure that the pressure difference $P_{ice} - P_{water}$ at both boundaries is determined by the temperature at the boundary location. The thermal regelation process is proposed to increase the effective hydraulic conductivity greatly in the frozen fringe, and if the hydraulic conductivity is high, relatively little suction is needed to move water. Hence, the pressure on the water in the frozen fringe need not go much below the pressure on the water in the unfrozen soil for continued inflow. In essence, the unfrozen soil water acts like a reservoir to help maintain P_{water}. But $P_{ice} - P_{water}$ is determined by the temperature, so maintenance of P_{water} drives P_{ice} upward as the temperature falls. Thus an important characteristic of the rigid-ice model is that P_{ice} is allowed to rise, rather than given a value equal to the overburden pressure as in the capillary and hydrodynamic models. When P_{ice} rises enough to equal the overburden pressure, the rigid ice lattice lifts the soil and allows a layer of segregation ice to form. The layer ceases to grow when enough desiccation takes place to lower P_{water}, and that brings P_{ice} below the overburden pressure, terminating the growth. The freezing front then advances until enough inflow of water makes P_{water} again rise, and the cycle repeats.

The premelting model of segregation ice formation

The *premelting* model is relatively new, having been developed primarily during the past decade. Like the capillary, hydrodynamic, and rigid-ice models, this model depends on examination of the free energy of a system, but it takes into account a component of the system's free energy that the others do not: that associated with a stable layer of water sandwiched in between ice and soil surfaces. As early as 1850 Michael Faraday and others were making observations that suggested to them that stable layers of water existed on snow or ice surfaces. A contemporary, highly influential Lord Kelvin, incorrectly thought that Faraday's observations were due simply to the melting of ice under pressure, so the existence and characteristics of the stable water layers did not receive the attention deserved until much later. Now, the existence of the stable layers is proven, and much is being learned about them.[30]

If a solid substance such as ice is in contact with its vapor phase when the absolute temperature of the solid is below approximately 0.9 its absolute melting temperature T_0 (for ice, $T_0 = 273$ K), then the interface remains dry; that is, no liquid phase is present. However when the temperature rises to approximately $0.9T_0$, then the ice system's free energy is lowered if some of the ice melts to form a layer of water between the ice and the vapor. At first, this so-called premelted layer is only one or two molecules thick, but the layer thickens as the temperature rises, moving toward macroscopically large thickness at the melting temperature. This change from a two-phase to three-phase system is of course a consequence of the Universal System Happiness Rule, and the rule also requires that the thickness of the premelted layer be that which minimizes the overall free energy of the system. Note that the premelted layer is a stable layer, whereas a layer of supercooled bulk water is only metastable—tickle it slightly by introducing a large enough solid surface and it will freeze. Thus

---

30. Dash et al. (1995).

supercooled bulk water depends on the absence of a solid surface, whereas premelted water requires one.

In the premelting model of segregation ice formation, a first step in the approach is to predict how the thickness of a premelted layer and the pressure within it vary with temperature of a system. The system considered is one consisting of a premelted layer separating an ice wall and a wall of another substance such as soil. As with other models discussed above, the approach is to examine the free energy of the system to ascertain the conditions required for the system to be in equilibrium; that is, to possess minimum free energy. However, the premelting examination goes farther by including in the accounting the free energies of all the interfaces along with the bulk free energies of the ice and the water plus the free energy associated with intermolecular forces acting between all the molecules in the three layers: ice, water, and soil.

The intermolecular forces of concern are Van der Waals forces and electrostatic forces that might involve distributions of surface charge at interfaces and ions within the premelted layer if it contains impurities. These various forces decay with distance in different ways, so the predicted thickness of the premelted layer depends on which type of force is incorporated in the computation. The procedure is to fix the temperature and pressure and then see what thickness of the premelted layer allows the system to have the least free energy.[31, 32, 33] The process leads to an expression for the premelted layer's thickness d:

$$d = \lambda_v T_0^v / (T_0 - T)^v \qquad (5)$$

where T is the temperature, T_0 is the freezing temperature of bulk water (0°C), and λ_v is a fixed coefficient. Equation 5 looks simple, but much is bound up in the coefficient λ_v. It depends on the heat of fusion of water (L), on the density of ice ρ_{ice}, and on v, and v, in turn, is a number related to the fashion in which the intermolecular Van der Waals or electrostatic forces decay with distance. For Van der Waals forces, $v = 3$ or 4, and for electrostatic forces $v = \frac{3}{2}$ or 2.

Also coming out of the examination of the free energy of the system is an expression relating the hydrodynamic pressure on water in the premelted

---

31. Wettlaufer et al. (1996).
32. Worster and Wettlaufer (1999).
33. Note to the mathematically inclined: The expression for free energy is expanded into a power series and then the first-order term is taken as the result.

layer (P_{water}) to the pressure on the ice (P_{ice}) and the temperature. The pressure in each of these phases is uniform, but the interactions between water and ice molecules create a difference between P_{ice} and P_{water}, and these interactions depend on the thickness of the premelted layer. The expression for the hydrodynamic pressure in the premelted layer (P_{water}) is independent of the nature of the intermolecular interactions; that is, it does not depend on the numerical value of v. The pressure in the water is dependent only on the specific density of ice ρ_{ice}, the heat of fusion L, and the temperature T:[34]

$$P_{water} = P_{ice} - \rho_{ice}L(T_o - T)/T_o \qquad (6)$$

Notice that this result differs from that given by Equation 2, derived from the Clausius-Clapeyron equation with P_{ice} held fixed. The appearance of the density of ice ρ_{ice} in Equation 6 but not in Equation 2 is the cause of the difference, and the consequence is that the predicted change of pressure on the water changes by approximately 11 Atm/°C instead of the 12Atm/°C predicted by fixing the pressure on the ice and solving for the pressure in the water using the Clausius-Clapeyron equation. Representing Equation 6 on Figures 3.10A and B are the lines labeled "Premelting Model." Curiously enough, the result is identical to that derived from the Clausius-Clapeyron equation with P_{water} instead of P_{ice} held fixed.

Notice from Equation 6 that the pressure on the ice P_{ice} is equal to the pressure on the water plus another pressure equal to $\rho_{ice}L(T_o - T)/T_o$. It is convenient to call this pressure the **thermomolecular pressure** P_T. In the premelting model it is assumed that the pressure on the ice in a freezing soil is initially fixed and equal to the overburden pressure. Fixing P_{ice} requires that P_{water} decrease and the thermomolecular pressure increase with decreasing temperature. Hence if there is a temperature gradient in the freezing soil, P_{water} and P_T vary with position, and the suction $P_{ice} - P_{water}$ draws water toward the coldest part. Not only is water pulled to the freezing front from the unfrozen soil, it also travels via the premelted layer into the frozen soil. However, the thickness of the premelted layer declines rapidly with temperature, less rapidly if electrostatic intermolecular forces dominate rather than Van der Waals forces, but in both cases the change with temperature is profound.[35, 36] The thickness at −1°C is 1/100 to 1/1000 that at −0.1°C, and at −10°C the thickness is 1/1000 to 1/1,000,000 that at −0.1°C.

34. Dash et al. (1995); Wettlaufer et al. (1996); Worster and Wettlaufer (1999a).
35. Wettlaufer (1999a).
36. Wettlaufer (1999b).

Thus pulling water through the premelted layer is like pulling water through a funnel with extreme taper. Near the relatively warm freezing front the channel is wide and water flows relatively easily, but where the temperature is lower, the channel narrows and so greater suction is required to maintain a given flow. Because the suction changes much more slowly than the thickness of the premelted layer with falling temperature, water will tend to pile up somewhere behind the freezing front. This incoming water increases P_{water} locally, and that in turn drives P_{ice} upward locally, causing it to exceed the overburden pressure locally, rupturing the soil. The increase in P_{water} concurrently raises the free energy of the water above that of ice at the temperature existing and so the water converts to segregation ice. Ice will continue to form as long as enough water flows to the forming layer to maintain P_{water} high enough to force P_{ice} to exceed the overburden pressure. If enough desiccation occurs then P_{water} falls, dragging P_{ice} down below the overburden pressure to terminate the growth of the layer of segregation ice. The freezing front advances, and when conditions again become favorable a new layer of segregation ice forms at a different location.

The premelting model allows very high suctions and heaving pressures to develop since the premelted layer is stable down to temperatures near −30°C, where the heaving pressure exceeds 300 Atm. That is equal to the pressure difference between the bottom and top of a water column 3400 m (11,200 ft) tall. That premelting can generate segregation ice has been demonstrated in the laboratory, in situations not involving curved interfaces so that capillary effects could be ruled out.[37]

Summary discussion of the models of segregation ice formation

Common to all the models put forth to explain the formation of segregation ice is the idea that temperature-dependent suction (cryosuction) develops in a porous freezing soil and that this suction draws water into the soil. The transported water freezes somewhere within the soil, typically in the form of layers perpendicular to the imposed gradient in temperature, and these increase the volume of the soil. Coexistence of water and ice in the freezing soil requires that the pressure in the water be different from that in the ice, and that these two quantities P_{water} and P_{ice} be related in a way that depends on temperature, namely growing farther

---

37. Wettlaufer (1999a).

apart as the temperature of the soil falls below 0°C, by 11 Atm/°C in the premelting model, and by 12 Atm/°C in the others that adopt a version of the Clausius-Clapeyron equation wherein the pressure on the ice is held fixed. Both predicted pressure differences (suctions) $P_{ice} - P_{water}$ fit the data at near-0°C temperatures, but the suction predicted by the premelting model fits better at lower temperatures. Since in all cases the suction $P_{ice} - P_{water}$ is constrained to a very narrow range, 11Atm/°C to 12 Atm/°C, the exact value of this suction is not important. It probably varies with change in the rate of water inflow from nearby unfrozen soil, and its variation might be a factor helping to determine the thickness of segregation ice's layers. All models are compatible with the idea that the only way to stop the flow of water into the soil, and therefore to avoid increasing the volume of the soil-ice matrix, is for the overburden pressure to exceed P_{ice} everywhere in the freezing soil.

Recall that the shapes of soil freezing characteristic curves are very similar to, and can be predicted from, soil moisture characteristic curves. That fact suggests that the small-scale intermolecular forces which cause water to cling to soil so tenaciously are much the same whether the soil be frozen or unfrozen. Ground temperatures rarely if ever get down to -30°C (approximately 0.9 times the absolute melting temperature of ice). Therefore, an interesting consequence of premelting is that the phenomenon guarantees that all or virtually all of the soil particles in the world's frozen ground, be they in the active layer or in permafrost, will have a coating of liquid water surrounding them. This coating provides an avenue along which water can travel whenever it is under suction.

▪ Frost Heave

Most geocryologists use the term "frost heave" to mean the increase in volume undergone by a soil because of the segregation ice forming within it. By this strict definition, any increase of a soil's volume due to the freezing of its resident pore water is excluded, but that is usually of minor consequence anyway, never exceeding 9% and typically far less. Much greater expansions do occur, and so it is to explain how these expansions come about and to predict how large they will be that geocryologists have developed the various models to explain the formation of segregation ice and its legacy, frost heave.

Such a delicate balance must occur among so many contributing factors that accurate prediction of frost heave is not yet within reach. Reviewing the key features of the models brings out several important requirements for the formation of segregation ice.

Segregation ice will form if:

1. Subzero temperature gradients and soil pore sizes permit cryosuction to develop, and
2. a source of water is available, and
3. the permeability (effective hydraulic conductivity) allows water transport to the site where the ice can form, and
4. the thermal conductivity and temperature gradient permit a balancing of heat energy in and out (including the latent heat of fusion)[38] sufficient to stall the freezing front or allow advance at a rate no faster than the segregation ice accumulates, and
5. the suction on the water in the soil is such that the conversion of water to ice allows the ice pressure to equal or exceed the pressure from the weight of overlying material.

The ubiquitous growth of pipkrake ice shows that these five necessary conditions are easily met at the surface of the ground when the air temperature falls only slightly below 0°C. The pipkrakes grow where the ground is wet and loose, so only minor cryosuction is required to bring water to the freezing front. Here, segregation ice readily accumulates because the lack of overburden minimizes downward pressure and permits easy removal to the air of the freezing water's heat of fusion.

Below the surface of the ground, things become more complicated. The observed development of multiple layers of segregation ice in freezing soil, both in the laboratory and in natural settings, hints strongly that the system cycles through values that generate the five necessary conditions and then cause them to hold no longer. In a given soil the two variables that can change are the temperature and the availability of water. Segregation ice forms in a soil having favorable characteristics (mainly pore size and effective hydraulic conductivity) when the temperature-driven suction delivers the proper amount of water to and beyond the freezing front.

---

38. The thermal conductivity and the effective heat capacity change abruptly at the freezing point because both depend on the water content, and that in turn depends on temperature. Ice is more conducting than water, so as more water converts to ice in a freezing soil its thermal conductivity increases. The effective heat capacity in the temperature range –3°C to 0°C is very high because in this range much change of phase takes place and the water's latent heat of fusion is so large.

The segregation ice can continue to form as long as the water supply is adequate, but the necessary suction tends to deplete water from the surroundings. If the supply of water becomes inadequate, the growth of segregation ice terminates and the freezing front advances. Later, and at some distance beyond, continuing temperature decline will increase suction and bring enough water to begin forming a new layer of segregation ice.

Multiple layers of segregation ice tend to increase in thickness with depth, as was shown in Figure 3.12. As the active layer freezes from the top down, the temperature changes more slowly with depth, and that gives the existing combination of suction, effective hydraulic conductivity and water supply a longer opportunity to maintain proper conditions for continuing growth of a segregation ice layer.

Gradations and boundaries in the soil help determine the location and shape of bodies of segregation ice. Thick accumulations are often found near boundaries between zones of fine-grained material and more permeable coarse-grained material that favors movement of water to the site where segregation ice is forming. Compositional variation of any sort that affects the effective hydraulic conductivity, thermal conductivity, or heat capacity can influence the location, orientation, and shape of ice layers. And any ice layer that forms tends to foster additional accumulation of segregation ice because the boundaries of the initial ice layer generally are locations of abrupt changes in effective hydraulic conductivity. That segregation ice does tend to build up at these sites is verified by laboratory observations of simultaneous enlargement of multiple ice layers within frozen soil.[39]

Recall from viewing Figure 3.15 that at temperatures barely below 0°C the effective hydraulic conductivity of ice is less than that of saturated frozen soil but that at lower temperature the effective hydraulic conductivity of the ice exceeds that of the soil. Thus, depending on the temperature, water molecules might encounter either increase or decrease in effective hydraulic conductivity as they move through frozen soil into an ice layer. If they encounter a decrease they should tend to pile up at the boundary, causing a growth of segregation ice there. If the temperature at the boundary causes the effective hydraulic conductivity of the ice to be greater than the hydraulic conductivity of the soil, then the water will tend to pile up at the exit boundary of the ice body. **Figure 3.17** indicates schematically on which sides of ice layers segregation ice should tend to accumulate during

---

39. Williams and Smith (1989) pp 232–33.

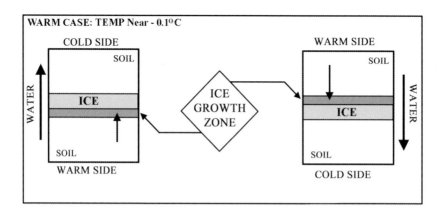

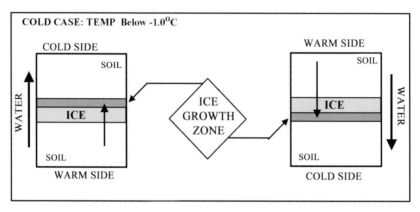

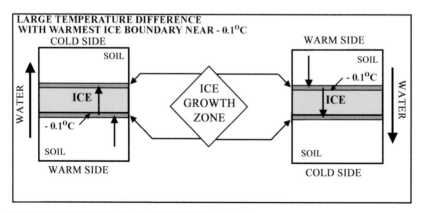

Figure 3.17 Top panel: When the soil and ice temperature is barely below freezing, and the ground surface is colder than at depth, water moves upward to create new ice on the bottom of a pre-existing ice layer (as at left), but if the temperature decreases downward, water moves down and new ice forms at the top of an older ice layer. Center panel: When the soil and ice temperature is well below freezing, just the opposite occurs: new ice accumulates at top during the fall and at the bottom during the spring. Bottom panel: If a large temperature difference appears across a pre-existing ice layer, then ice can form on both sides of it, fall or spring.

freeze and thaw periods when the temperature at their boundaries is just below and well below freezing.

That part of the permafrost layer lying above the level of zero amplitude (see **Figure 3.18**) in the annual temperature variation undergoes temperature fluctuations ranging between 0°C (at the very top of the permafrost) down to the minimum temperature of the permafrost, about −10°C in northern Alaska and −13°C in northeastern Siberia. At or somewhere below the level of zero amplitude, typically at depth 10 to 25 m, the normal upward geothermal heat flow causes increasing temperature (the geothermal gradient ranging from 1°C/20 m to 1°C/60 m [1°F/40 ft to 1°F/110 ft]). Thus, wherever the top of the permafrost lies above the zero-amplitude level, a continuing growth of ice layers can occur. In winter cryosuction pulls water upward through the permafrost and in summer it helps drive the water down from the thawing active layer above. So, as in a tidal channel, though much more slowly, water sweeps back and forth through the top of the permafrost with little volume, and over distances perhaps measured only in centimeters. Still, over the course of years the flow can cause—millimeter by millimeter—growing accumulations to any horizontal ice layer already in place.

As long as the ice layers are thin, at any time they are likely to accumulate segregation ice on one side or the other, as indicated in the top panel of Figure 3.17, but the accumulation might shift from one side to the other as the temperature changes from near zero to somewhat colder (compare left-hand or right-hand parts of the top and center panels in Figure 3.17). Then if the ice layers become thick enough for the temperature gradient to produce substantial temperature differences across the layers, simultaneous accumulation can occur, as in the bottom panel of Figure 3.17.

The long-term consequence of the *cryogenic* pumping of water through frozen ground is an enrichment of the ice content of the uppermost part of the permafrost layer and the formation of layers and lenses of pure ice near the top of the permafrost, either singly or in association with other forms of ground ice, such as shown in **Figure 3.19**. Geocryologists sometimes refer to the new segregation ice forming within previously frozen ground or that accumulating to pre-existing ice layers as *aggradational ice*. In some instances it becomes very thick, commonly in excess of 10 m in the Mackenzie delta of northwestern Canada, and sometimes more than 30 m.

Based on field observations, laboratory experiments, and interpretations guided by the models of ice segregation, it is clear that key factors

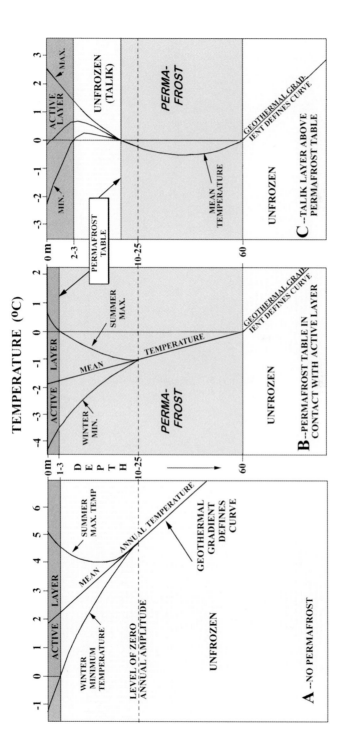

Figure 3.18 Schematic plots of ground temperature where : A—no permafrost is present below the active layer, B—the permafrost table is in contact with the bottom of the active layer, and C—the nonequilbrium situation where climatic change or other cause has melted the upper portion of the permafrost to leave an unfrozen layer (talik) between the bottom of the active layer and the permafrost table.

Figure 3.19 Ice wedges and multiple layers of segregation ice in a roadcut near Fox, Alaska. See also Plate 6.

determining the rate and magnitude of ice growth—and hence of frost heave—are water supply, soil grain size, and the thermal regime. The low permeability of clay limits water supply, and so the largest frost heaving is in soil with silt-sized particles. Large-grained soils—sands, gravels, and also peats—support little if any frost heave because they do not allow large suctions to develop. In most instances when the temperature falls rapidly (and hence the associated heat flow rate is high) the rapid advance of the freezing front, at least at shallow depths, limits segregation ice to thin layers, and so the heave also is small. However, even in that situation of rapid change, the heave can be very large if the supply of water is copious and the soil permeable. Then enough water may migrate to and freeze near the freezing front to stall the advance of the front by giving up its heat of fusion and thereby forming a very thick ice layer that rapidly heaves the ground surface upward.

When the ground first freezes, the rate of heaving due to forming segregation ice is small, but it increases to a maximum when conditions cause the freezing front to become nearly stationary. Thereafter, as shown in **Figure 3.20**, the rate of primary heaving falls away as the rate of heat removal and the rate of water transport to the front decline. This fast initial ice formation and heaving presumably occur at temperatures only slightly below 0°C since they are restricted to a location at or very near the freezing front. The near-zero temperature there limits the ice pressure P_{ice} (the heaving pressure) because the difference in ice and water pressure $P_{ice} - P_{water}$ is also near zero, and the water pressure P_{water} will always have some finite value, typically one less than the pressure from the weight of overlying soil. If freezing is occurring at temperature $-0.1°C$, $P_{ice} - P_{water} = 1.1 - 1.2$ Atm, and if the water pressure is 1 Atm or less then the heaving pressure cannot exceed 2.2–2.4 Atm. That is the pressure exerted by a soil overburden layer approximately 6 to 10 meters thick (approximately 10 metric tons/m^2 or 2000 lbs/ft^2). Thus, the

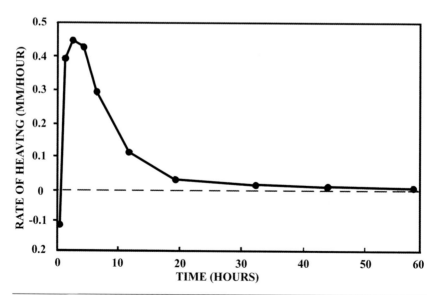

Figure 3.20 When a wet, fine-grained soil freezes, the rate of heaving climbs quickly to a peak and then declines. Redrawn after a diagram by Garand (1981)) reproduced as Figure 6.8 in Smith (1985a).

initial heaving—sometimes called the **primary frost heaving**—can readily lift the uppermost layer of soil and rather substantial objects placed upon it such as houses, roadways, and railroads.

If the overburden load is great enough it can stop the forming of segregation ice, and hence, frost heave. Then any further freezing forces the unfrozen water into the ground below, and if the soil there is very fine-grained the required pressure is very large—hundreds of atmospheres. Pressure-measuring devices placed in freezing soils have shown that such large pressures do occur, and the fact that the formation of segregation ice obviously has taken place at depths where the overburden load is in the vicinity of 20 Atm adds further verification. The initial (primary) frost heaving—that developing at or very near the freezing front where the temperature is near 0°C—is incapable of producing large pressures, and so in explaining these the rigid-ice and premelting models of ice segregation come into their own. They allow for ice formation behind the freezing front where the temperature can be several degrees below freezing and the ice pressure high. In keeping with the identification of the near-0°C ice segregation as causing primary heaving, some geocryologists identify the slower but more powerful expansion as **secondary frost heaving**. Whereas primary heaving probably occurs mainly within the annual freezing of the active layer, secondary heaving can extend down into permafrost to depths of 30 or more meters. The amount of segregation ice (and heaving) tends to decline with depth, and so most of the overall heaving is usually within the upper few meters of the permafrost layer.

Figure 3.21 presents a classic photograph, well illustrating both how large frost heave can be and how much it depends on local conditions, especially water supply. Although frost heaving of a few centimeters per year is typical, one set of bridge pilings in Siberia heaved 0.6 m (2 ft) in one year, and the piles at the right-hand side of the bridge in Figure 3.20 heaved 3.3 m (11 ft) in one year.[40] Railroads are particularly susceptible to problems with frost heaving since the rails need to be kept fairly level in order to avoid train cars becoming decoupled or derailed. The Alaska Railroad has found it necessary to expend funds each year for shimming up track, trimming off uplifted pilings, and replacing them. One cold day during the winter of 1950–51 the rear part of a passenger train came uncoupled while moving over a frost-heaved bridge just outside Fairbanks. The passengers sat cooling

---

40. Péwé (1982).

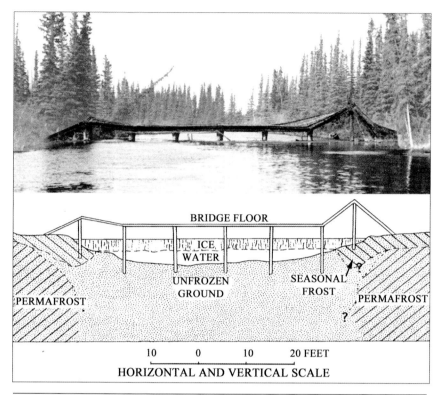

Figure 3.21 Frost-heaved bridge spanning the outlet of Clearwater Lake near Big Delta, Alaska. Compiled from a photograph taken by M. F. Meiser in August 1951 and a drawing by Troy L. Péwé, reproduced as Figures 30 and 31 by Oscar J. Ferrians et al. (1969).

their heels and all other parts of their anatomies for several hours while the engineer drove on the 80 km to Nenana, discovered his problem, and backtracked to the bridge raised by segregation ice.

▪ Other Kinds of Ice Found in the Ground

Frost heaving is generally considered to be the result of volume change wrought by segregation ice, that resulting from the freezing of water transported by cryogenic processes. Several other kinds of ice form in the soil, and also may cause uplift of the ground surface. One descriptive scheme geocryologists use has five categories: 1) segregation ice, 2) pore ice, 3) ice-wedge ice or foliated ice, 4) pingo ice, also called intrusive ice or injection ice, and 5) buried ice.[41] These various forms of ground ice may develop in association and even grade into one another, so sometimes it may be difficult to determine which kind is involved in a particular deposit.

Pore ice

Pore ice is that formed from water already in place within the soil. Any expansion of the soil that may result because of ice's 9% greater volume than water is in addition to cryogenic frost heave. Even if the freezing of pore ice causes little or no overall expansion of the soil, it may temporarily or permanently modify the soil by squeezing or otherwise altering the soil particles. Pore ice comprises a substantial part of the upper few meters of permafrost, and the ice in permafrost below depths of 10 m is mostly pore ice.

Ice-wedge ice or foliated ice

Ice-wedge ice is that formed in permafrost from surface water running down into small (1 to 3 cm) cracks opened by thermal contraction of the soil. The process is repetitive, progressing over the course of many summer thaws and winter freezes, so ice-wedge ice typically is highly foliated, each layer representing a freeze-thaw episode.

Pingo ice

Pingo ice (also called intrusive ice or injection ice) is that which forms from water under hydrostatic pressure in permafrost areas. It may be milky from entrapped air bubbles but is typically clear or nearly clear and is often found in planoconvex masses that, when forming, have lifted overlying soil upwards to create minor mound structures such as shown in **Figure 3.22** or the much larger mounds called pingos, discussed in Chapter 4.

Buried ice

Depositional events such as avalanches, mudslides, or windstorms sometimes bury sea ice, lake ice, river ice, overflow ice (aufeis), glacier ice, and

---

41. Péwé (1975b).

Figure 3.22 A layer of pure ice pushing up the moss in northern Alaska. Photograph by Richard Veazey. See also Plate 7.

recrystallized snow. This buried ice does not directly lead to frost heaving, but it might contribute to it by affecting where subsequent segregation (or aggradation) ice forms.

Another icelike substance: gas hydrates (clathrates)

Gas hydrates are icelike crystalline solids formed from methane gas and water molecules. In one kind of gas hydrate (Structure I), 46 water molecules join to form a large lattice structure in which some methane molecules (CH_4) participate, and in another (Structure II) 136 water molecules form a lattice in which ethane molecules (C_2H_6) can join. The maximum participation allows one methane molecule in Structure I for every $5\frac{3}{4}$ water molecules, and one ethane molecule for every 17 water molecules in Structure II. This packing brings the methane and ethane molecules into a volume 200 times smaller than they would occupy if by themselves under the same conditions of pressure and temperature.[42] It is as if the gas molecules were locked into a water molecule pressure vessel. However, the vessel

42. Ehlig-Economides (1981).

is stable only in a limited range of pressure and temperature conditions. At pressure 20 Atm the gas hydrate is stable only at temperatures below −10°C, at 10 Atm it is stable only below −30°C, and at 1 Atm stable only below about −80°C. The necessary conditions for stability exist in the coldest of permafrost areas at depths below 100 m and perhaps down to a depth of approximately 1000 m where heat flow from the earth's interior raises the temperature above the maximum for stability at the pressures involved. Conditions for stability should also pertain beneath the oceans, at depths near 250 m in cold arctic waters, and at depths below 400 m in warmer waters.[43]

Drilling into gas hydrates can be hazardous because the release of pressure on the hydrates can cause an almost explosive expansion that can blow drilling equipment back up the hole. However, the hydrate deposits in and below permafrost also represent a potentially huge energy resource.[44]

▪ Chapter Summary

Various kinds of suction—vacuum suction, capillary suction, osmotic suction, and cryosuction—act to move water from one place to another because the movement lowers the overall free energy of the system. Cryosuction, the suction that develops in the pores or on the walls of fine-grained materials at subfreezing temperatures, is a powerful water-moving force. A familiar example of its operation is the freezer burn that occurs when a poorly wrapped slab of meat is placed in a freezer. The cryosuction pulls the water out of the meat and converts it to ice crystals on the meat's surface or elsewhere in the freezer compartment. The same thing happens to the soil near the surface of the earth when the air temperature goes down below 0°C. Cryosuction in the soil pulls water toward the cold surface where it often forms ice crystals called pipkrakes or needle ice. Cryogenic suction depends on both temperature and soil characteristics, particularly the size of the particles in the soil. Because of water's abnormal characteristics, cryosuction increases dramatically as the temperature falls, by 11 to 12 Atm per centigrade degree below 0°C, and in a fine-grained soil, it can exceed 100 Atm. An important consequence of cryosuction is the formation of microscopic to thick segregated layers of pure ice within fine-grained soils when adequate water is available and the

43. Williams and Smith (1989) pp 46–48.
44. Ehlig-Economides (1981).

thermal regime is favorable. The formation of this segregation ice can increase the volume of a soil by 50% or more, and in the process heave the ground surface upward. Silty soils are the most prone to frost heave, and in them the heaving is powerful enough to lift the uppermost layer of soil and structures built thereon. Segregation ice is widespread, and it forms in both seasonally and perennially frozen ground. Of the mechanisms responsible for forming the various types of ice found in the ground—segregation ice, pore ice, foliated ice-wedge ice, pingo ice, and buried ice—the one creating segregation ice is the most complex, and also the least understood.

Land Forms Created
by Cryogenic Action

■ Introduction

Cryogenic action in the active layer and in permanently frozen ground produces a variety of geomorphic features unique to cold lands, and also a few that appear very similar to landforms created by geological processes not related to freezing and thawing. Some of the cryogenically caused landforms and effects are relics of past times when the climate was colder; many others are the consequence of ongoing processes that wax and wane with variation in weather and climate. Cryogenic processes continuously operate to modify the land surface and the ground below wherever the temperature of the ground dips below freezing, even if just for a few hours. Some changes occur rapidly, in a matter of hours or days, and others proceed slowly, over the course of thousands of years. But be they fast or slow, the changes wrought by the freezing and thawing of the ground lend special interest to the overall topic of perennially and annually frozen ground.

The formation of pipkrakes described in Chapter 3 is an example of a fast-acting and readily observed cryogenic process that operates over much of the earth's land surface, and it provides obvious demonstration of the profound effect of cryosuction on movement of water through the ground—typically in the direction toward the cold. That movement is crucial to the formation of most if not all transient and long-lasting landforms generated by freezing and thawing of the ground. As we now explore their distribution and characteristics it is good to keep in mind that the various kinds of landforms, like the different kinds of ground ice, may intermingle, and so what a person sees in any one place may be the result of several generative processes. Also it may be comforting to keep in mind that the names attached to these various landforms are important only to the extent that they may be useful to the discussion of the formative processes.

■ Ice Wedges and Polygons

Ice-wedge arrays

Ice-wedge arrays are widespread in permafrost, and one geocryologist has estimated that *ice wedges* comprise a volume that is 10% of the top few meters of permafrost on the northern coastal plain of Alaska.[1] In that region and in many other areas the polygonal pattern in which ice wedges form expresses itself on the ground surface so that it is readily recognized, especially when seen from the air. Ice wedges also can lie unseen in the permafrost—down beneath an insulating cover of soil, moss, and trees—capable of destroying the works of the unsuspecting engineer or homeowner who disturb its environment by placing a heated structure overhead or removing the overlying insulative cover.

The generally accepted explanation of how ice wedges form is a simple one: a wedge begins when water freezes in a crack opened in the ground when it contracts from the cold.[2] The ice-filled crack is weaker than the surrounding frozen soil so it breaks open again when the ground next cools enough to cause cracking. Thus the process cycles on year after year to build a tapered wedge of ice up to several meters wide at the top and perhaps five or more meters tall. Each summer, warming causes expansion of the frozen soil adjacent to the wedge, and it undergoes plastic flow that bends it upward. In some places the wedges are so extensive that they no doubt have elevated the overall ground surface. Classic photographs of an initial crack and of a well-developed ice wedge appear in **Figures 4.1** and **4.2**, and **Figure 4.3A** shows a schematic drawing of how the wedges develop.

---

1. Brown (1967); Péwé (1975b).
2. Leffingwell (1919); Lachenbruch (1962).

Figure 4.1 A frost crack on the surface of a recently drained area on Flaxman Island. From Plate XXIX in Leffingwell (1919). Cracks such as this occur abruptly, generating loud noises and very small, localized earth tremors called cryoseisms by Lacroix (1980).

Spacing and orientation of contraction cracking

Polygonal cracking is a commonly observed phenomenon in materials that contract when cooled or desiccated.[3] Examples of cracking caused

---

3. Any process that causes the volume of a material to decrease in a way that creates internal tension may generate contraction-crack polygons, so phase changes and chemical reactions might also generate the cracks.

Figure 4.2 Foliated ice wedge in silt exposed by placer mining operations near Livengood, Alaska. Photograph No. 474 by Troy L. Péwé, September 1949.

by cooling are the crackling in paint on metal surfaces, the crackle finish on some pottery, and shrinkage cracking of concrete and of some basaltic rocks. Commonly seen examples of desiccation cracks are those appearing in mud when it dries. Since the cracks occur because the

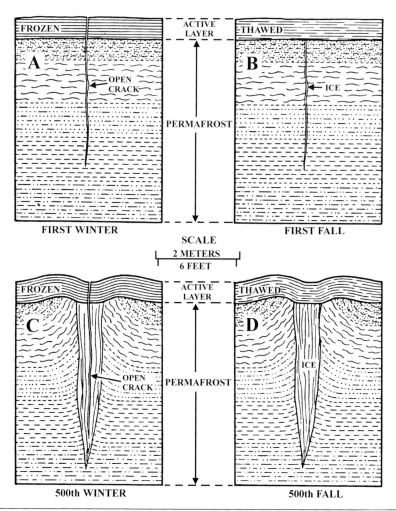

Figure 4.3A Schematic representation of the evolution of an ice wedge according to the contraction-crack theory. The crack is exaggerated for illustrative purposes. Diagram modified slightly from Figure 1 of Lachenbruch (1962).

material has failed under tension, the various kinds of cracking share common characteristics. The scale of the cracking differs because it depends on the tensile strength of the material involved, and the cracks are less uniform when the cracked material is not homogeneous. Certainly one of the least homogenous materials is soil, so it follows that when the ground undergoes cracking during cooling, the cracks are likely to be far from perfectly uniform.

Frozen ground gains tensile strength and becomes increasingly brittle as it grows colder. However, rapidly falling temperature causes differential shrinking in frozen ground and thereby generates high enough internal *stress* that exceeds the tensile strength of the frozen ground and causes it to crack. The cracking involved in the development of ice wedges typically occurs in midwinter when the near-surface ground temperature is –15°C or below and rapidly falling air temperature further cools the ground by another 10 or 20 degrees. Ice wedges become widespread where the mean annual temperature is lower than –7°C. The cracking happens abruptly, and people report hearing rifle-shot noises as the cracks break open to widths ranging from a few millimeters to 2 cm at the top and tapering down with depth in consequence of lowered stress there.

Geocryologists believe that the initial cracking that begins the development of a polygonal array of ice wedges starts at the ground surface and propagates downward into the active layer and permafrost below until compressive forces created by overburden help the frozen soil's inherent strength stop further cracking. In permafrost areas the final depth of cracking typically is 3 to 10 meters. That initial cracking relieves stress in the direction at right angles to the plane of the crack (by convention, called the normal direction). All the normal stress just beside the crack disappears, and the amount of normal direction stress release declines with distance out away from the crack. For cracks 3 to 10 meters deep the normal stress is down to about one-third the initial value at a horizontal distance of 3 meters, and it falls to 5% at distances of 10 to 20 meters. That distance—10 to 20 meters—is roughly equal to the diameters of *polygons* that develop when further cracking occurs. A factor determining the polygon size is how rapidly the ground cools while the cracks are forming; the more rapid the cooling, the smaller the polygon.

Contraction crack polygons generally fall into two main categories called orthogonal and nonorthogonal polygons. Orthogonal polygons tend to have four sides, since in orthogonal cracking the cracks tend to meet at right angles. The nonorthogonal polygons tend to have six sides since the responsible cracks tend to meet at angles of 120°, as shown in **Figure 4.3B**.

The Universal System Happiness Rule dictates that highly homogeneous materials should undergo nonorthogonal cracking because that form releases the most strain energy per unit crack area and therefore leads to a configuration of least free energy.[4] Because naturally occurring materials are rarely homogeneous, nonorthogonal cracking is rare, but a

---

4. Lachenbruch (1962) p 49.

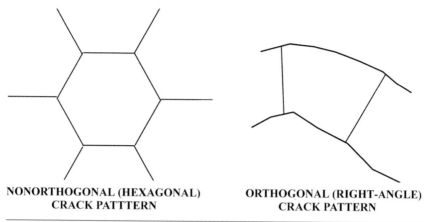

NONORTHOGONAL (HEXAGONAL)   ORTHOGONAL (RIGHT-ANGLE)
CRACK PATTTERN   CRACK PATTERN

Figure 4.3B Schematic drawing showing the nonorthogonal cracking typical of homogeneous materials, and the orthogonal cracking typical of materials such as soil that have nonuniform characteristics.

familiar example is the formation of hexagonal columnar blocks in cooling lavas. Another seen on occasion is hexagonal cracking in thin layers of drying moss or other humus matter atop the ground. Orthogonal cracking predominates in nature because of the inhomogeneity of the materials involved. Desiccation cracks in drying mud typically are orthogonal, and most permafrost polygons are also.

Cracking commences in frozen ground where the material has flaws and also where horizontal temperature gradients exist. For the latter reason the first cracking is likely to occur parallel to the shores of lakes or streams. Secondary, typically orthogonal, cracks then develop until the eventual result is a polygonal array of cracks that can evolve into ice wedges if filled with water that freezes. If the initial cracking leading to a polygonal array is somewhat curved, favored sites for secondary orthogonal cracking are on the convex side of bends, such as those shown in **Figure 4.4.**

Growth of Ice Wedges

Although the first-year cracking that initiates the formation of ice wedges most likely begins at the ground surface, the cracking during subsequent years probably begins lower down, at the top of the permafrost.[5] Obviously,

5. Lachenbruch (1962).

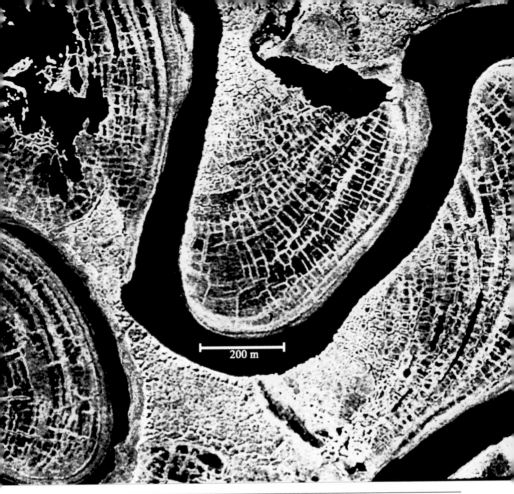

Figure 4.4 High-altitude infrared photograph (false color) of ice-wedge polygons on Alaska's North Slope. Courtesy Geophysical Institute, Fairbanks, Alaska. See also Plate 8.

any ice that might form in cracks within the active layer during early spring would melt in summer, allowing the cracks to slump or reseal through expansion of the warming soil. But essential to the continuing development of ice wedges is that the ice-filled cracks formed in the permafrost are sufficiently weak to break again before the soil cracks during the next strong cycle of thermal contraction. Renewed cracking at the center of each growing ice wedge and subsequent freezing of added water may occur on average only one out of two years, but after many years of repeated cracking an ice wedge can develop to a thickness exceeding 2 m.[6]

6. That a crack once formed will break again when the conditions are right would seem to be expected in view of the premelting phenomenon since it requires the existence of a water film at the interface whenever the temperature is warmer than approximately −27°C.

A distinguishing characteristic of ice-wedge ice is its foliated structure, the consequence of the method of formation. Each foliation layer represents one annual growth increment and is marked by air bubbles or other inclusions such as soil particles. The foliations show readily in Figures 4.2, 4.9, and 4.10, where the striated appearance is the two-dimensional manifestation of the three-dimensional foliation.

Most of the water coming into the central crack of an ice wedge undoubtedly runs down from the active layer above, but it is possible for water to move horizontally or upward through the wedge ice under the influence of a temperature gradient. At temperatures fractionally below 0°C both the thermal and hydraulic conductivities of ice exceed that of frozen silty soil, so an inward horizontal migration of water toward the center of the wedge can occur when the ground is cooling in the fall. Similarly, when the ground is warming, the migration could reverse to cause an outward transport of water through the ice. Such cryogenic transport raises the possibility of new ice forming near the boundaries of an ice wedge as well as in the core. However, the observed orientation of ice crystals—parallel to the boundary rather than normal to it as in the core of the wedge—does not hint that this horizontal transport contributes to the growth of ice wedges. Also, the observed tendency for ice crystals to increase in size outward from the center of ice wedges indicates that the outer ice is the oldest, because the size of individual crystals in an ice body typically increases with age. In ice wedges, the size of the crystals ranges from 0.1 mm to 10 cm.[7] Nevertheless, it seems unlikely that all the ice in any ice wedge arrived at its location strictly by the flow of water under the influence of gravity during summer, since gradients in temperature within the forming ice wedges and the adjacent frozen soil—as well as in the active layer above both—are normally present. The consequent gradients in free energy of liquid water (the cryosuction) are bound to move some water, and all directions are possible, sideways as well as up or down. Also, because of premelting, the central crack in each wedge contains a liquid layer along which water can move if under cryosuction.

Active and inactive ice wedges, ice-wedge casts, and sand wedges

Various events can terminate the growth of ice wedges or lead to their demise. A warming climate might elevate winter temperatures sufficiently

---

7. Black (1974, 1978); Washburn (1980) p 49.

to curtail growth by stopping further cracking. In the same vein, anything—including landslides and burial by loess—that increases the insulation over ice wedges will lessen the rapid temperature changes that foster repeated cracking within them. If the growth of ice wedges has pushed up soil on either side, the resulting relative depression of the ground surface over the wedges can collect snow which acts as a good insulator. Also, an increase in moss cover might slow or terminate further growth of the wedge. The relief of stresses by fracturing within nearby ice wedges may allow some wedges to grow at the expense of others, or perhaps cause all to slow their growth.

Actively growing ice wedges typically produce *low-center polygons* on the ground surface because summertime thermal expansion pushs up soil adjacent to each ice wedge. Between the resulting two upthrust ridges a slight furrow may be evident; it lies directly over the top of each wedge. Thus evolves a low-center polygon, one with a central region lower than its boundary area. If an ice-wedge array becomes inactive and its top portion melts due to warming climate or removal of overlying insulating material, the resulting depression directly over the ice wedges leads to *high-center polygons*. Thus the topographic appearance of polygon arrays can indicate their degree of activity. As shown in **Figure 4.5**, the region where ice wedges are active in Alaska corresponds to the region of continuous permafrost (where mean annual temperature is about –7°C or lower, and the winter temperature at the top of the permafrost is –15°C or below). There, active ice wedges occur in silt, sand, and gravel, and low-center polygons are far more common than high-center ones. Farther south, in the colder part of the discontinuous permafrost region, the ice wedges are either weakly active or inactive, and it is usual to find more high-center polygons, especially in areas where the vegetative mantle has been removed. See **Figure 4.6**.

The shapes of ice wedges, their relation to each other and to the present location of the ground surface serve as indicators of past climate and past episodes of soil erosion or deposition. Climatic warming or soil erosion can lower the permafrost table from previous levels and thereby destroy or clip off the top of ice wedges previously formed (see **Figure 4.7**). Episodes of soil deposition can engender complex wedges similar to that illustrated in Figure 4.7C. Figure 4.7D illustrates a situation observed in a tunnel bored into permafrost at Fox, Alaska, just north of Fairbanks, (**Figures 4.8–4.10**) and in a nearby roadcut (**Figures 4.11–4.13**). In the tunnel one inactive array of wedges lies about 13 m below the present ground surface, and a second group of inactive wedges stands well above that, at depths of 2 to 10 m below the surface. Carbon-14 dating indicates that the lower array in the tunnel is about 32,000 years old, and

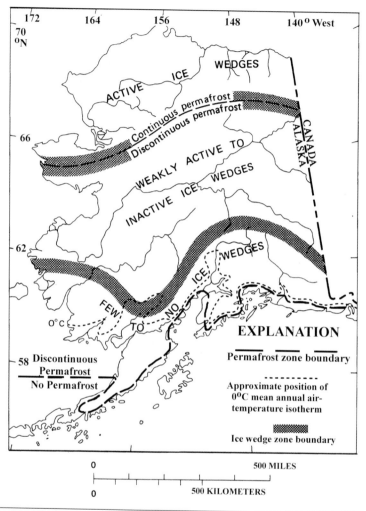

Figure 4.5 Distribution of ice wedges and permafrost in Alaska. Modified slightly from Figure 28 by Péwé (1975a)

that the upper and smaller wedges are 8,000 to 10,000 years old.[8] Evidently sometime between the formation of these two groups of ice wedges climatic warming and much new soil deposition occurred. The lower ice wedge array appears to have formed mostly during an epoch of minimal if

---

8. Sellmann (1967, 1972) cited by Péwé (1975a).

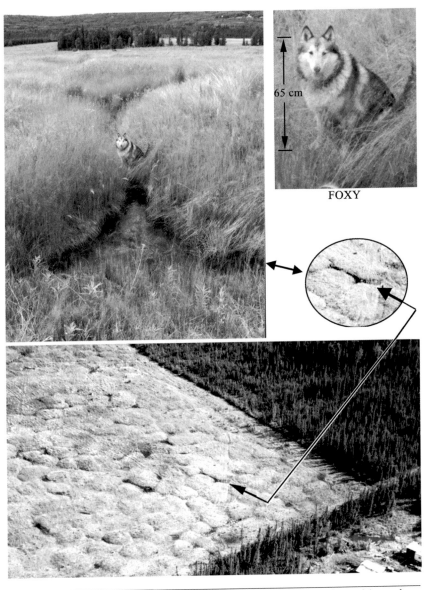

65 cm

FOXY

Figure 4.6A Top—Photo of thermokarst topography (high-center polygons) in an abandoned field taken in 1996, two days after the aerial photograph shown below. Microrelief in the thermokarst area is near 1 meter. On the higher ground such severe cracks parallel the depressions that it would be unwise to ride a horse across this field, located near the eastern end of the Farmers Loop permafrost area near Fairbanks. The scale dog seen in this and following illustrations is Foxy, a sheltie-husky cross 1 m long from tip of nose to base of tail. She stands 50 cm tall at the shoulder, and when sitting the top of her head is about 65 cm above ground. See also Plate 9.

Figure 4.6B Players at the North Star Golf Club—self-billed as the world's northernmost golf course—have their ups and downs as they putt through the high-center polygons. The fairways cover the right-hand side of the aerial photo taken at the same time as that in Figure 4.6A. The group of trees surrounding the green at top is the same as shown in the top part of the upper photo in Figure 4.6A.

any deposition. Geocryologists term such wedges (and also permafrost) as **epigenetic** to distinguish them from the so-called **syngenetic** wedges (and permafrost) formed during epochs of deposition. The upper grouping of ice wedges in the Fox permafrost tunnel appear to be of the syngenetic type. Syngenetic wedges, especially, may be quite complex in form. An example of what is probably a complex syngenetic wedge is shown in part B of **Figure 4.14.** The exposure shown in part A results from recent mining operations near Ester, Alaska, that have provided an unusual opportunity to observe the three-dimensional configuration of an ice wedge array.

Even ice wedges completely destroyed by thawing can reveal information about the past because soil spilling down into the voids left by melting ice wedges can form **ice-wedge casts** that become, in a sense, fossil ice

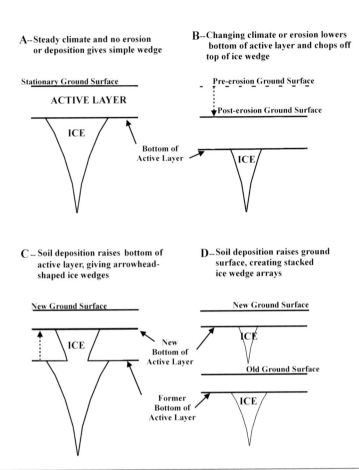

Figure 4.7 Schematic illustration of how: A) unchanging conditions can lead to classic-shaped ice wedges, B) erosion or warming can remove the upper parts of wedges, C) deposition can create syngenetic wedges with odd shapes, and D) extensive deposition after wedges form can lead to stacked arrays of wedges.

wedges.[9] However, not all soil-filled wedge shapes left by polygonal contraction cracking are ice-wedge casts because blowing sand or silt instead of water sometimes fills the contraction cracks. These similarly tapered deposits are called *sand wedges* to distinguish them from ice-wedge casts, and they differ also by having vertically oriented layers instead of primarily horizontal ones, as is depicted in **Figure 4.15**. Although moisture-rich, fine-grained soils are conducive to the formation of ice wedges in permafrost, those soils tend to flow so much that they inhibit the preservation of ice-wedge casts. Consequently most ice-wedge casts found are in gravel or other soil that can maintain steep faces when thawed.

9. Washburn (1980) pp 111–17.

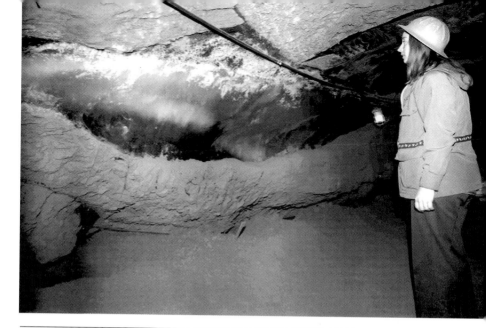

Figure 4.8 A thick layer of segregation ice or pond ice in the Fox permafrost tunnel. Its nonfoliated structure contrasts sharply with the foliated wedge ice shown in Figures 4.9 and 4.10. Photographed 1979; Fran Pedersen of the Geophysical Institute at right.

Figure 4.9A Foliated ice wedge cut by the Fox tunnel so that the remaining portion forms part of one wall and the ceiling of the tunnel. Some ablation of the surface has occurred subsequent to the boring of the tunnel in the early to mid-1960s. Photographed 1978; Rosemarie Davis at right.

Figure 4.9B Top part of an ice wedge overlain by silt containing massive, irregular-shaped and interconnected masses of segregation ice. Photographed 1979.

Figure 4.10 Intersecting ice wedges in the roof and wall of the Fox permafrost tunnel. Paula Jones and Fran Pederson of the Geophysical Institute stand directly below the intersection. Photographed 1979.

Figure 4.11 Ice wedges interconnected by multiple sheets of segregation ice in a roadcut approximated 1 km south of the Fox permafrost tunnel; photographed in 1979.

Figure 4.12A Part of the same roadcut shown in Figure 4.11.

Figure 4.12B Detail view of the ice wedge and interconnected sheets of segregation ice appearing in Part A just above and to the front of the truck cab. The configuration suggests that this is a syngenetic ice wedge, its upper part likely to have grown thousands of years later than the lower part that joins the layers of segregation ice.

Figure 4.13 Another view of the roadcut near the Fox permafrost tunnel which portrays the extensive nature of the multiple segregation ice layers interconnecting with ice wedges.

8 m

Figure 4.14A A polygonal array of ice wedges partly excavated by mining activities near Ester, Alaska. A large tractor has removed the tops of the wedges shown in the foreground and exposed cross-sections of those in the background. The interconnected dots show observed or inferred centers of the ice wedges in the array.

Figure 4.14B Close-up view of the complex ice wedge at upper left in Part A. (Photographed 1996.)

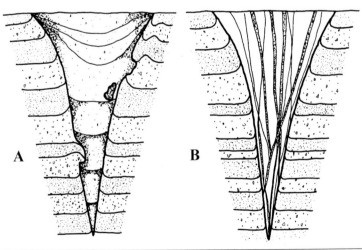

Figure 4.15 A) Schematic representation of a fossil ice wedge. It contains soil with slump structures but mainly displays a horizontal fabric. B) A sand wedge formed by sand falling into contraction cracks leads to a cast with vertical fabric. Reproduced from Figures 4.34 and 4.35 in Washburn (1980), originally drawn by R. F. Black.

Because ice wedges can form only in permafrost, their casts are one of the accepted indicators of former permafrost. Ice-wedge casts dating back more than 1 million years—to the earliest part of the **Pleistocene epoch**— have been found in Alaska, and many of Wisconsinan age (10,000 to 100,000 years B.P.) are found in other parts of North America where permafrost no longer is present.[10]

▪ Pingos, Palsas, and Other Protuberances

Pingos

"Pingo" is an Eskimo word for mound or hill, and, according to the teachings of former University of Alaska anthropologist Ivar Skarland, a much longer word starting with "pingo" means mound-that-blows-up. He said that the Eskimos gave this name to certain very strange large mounds that had broken tops and which therefore must have exploded. Skarland, suggesting it was merely legend, related a story he had heard about a man who, once, a long time ago, was standing on such a mound when it blew apart and threw him unhurt onto the nearby tundra. However, this event may actually have happened. Many of the ice-cored mounds now called pingos do have splits or depressions in their tops, caused either by melting of the top of the mound or a splitting open by dilatation (i.e., expansion) cracking in consequence of high hydraulic pressure within the pingo. Furthermore, two Russian scientists actually have observed the explosive dilatation rupturing of a pingo. It threw blocks of ice having volumes of 2 m³ into the air and cast smaller blocks as far as 22 m. The pingo simultaneously spouted a jet of water that lasted 30 minutes.[11] In other instances scientists have observed the escape of both water and nonexplosive gases.[12]

Pingos are large circular or elongated mounds having massive ice cores that have grown from water transported to the site by hydrostatic pressure. As the cores formed they pushed overburden upward as much as 50 m to create the pingo mounds. Sometimes looking like small volcanoes, pingos are perhaps the most exotic landforms created by cryogenic processes. As would any prominence rising discordantly above an otherwise swampy countryside, a pingo attracts attention. They are favorite nesting sites for small mammals such as foxes, and people also find them

---

10. Hopkins (1972) cited by Péwé (1975a).
11. Bogomolov and Sklyarevskaya (1973) cited by Washburn (1980) p 186.
12. MacKay and Steger (1966) cited by Washburn (1980) p 186.

appealing. Residents of Tuktoyaktuk in northern Canada have used the ice core of one pingo as a cold-storage facility, and in the vicinity of Fairbanks, Alaska, several people have purposely built homes atop pingos, sometimes without realizing the problems they likely would soon encounter.

Pingos are of two kinds, *closed-system* and *open-system*, referring to the nature of the water supply involved in the pingo formation. **Figure 4.16** illustrates the mechanisms believed responsible for the two pingo types. As is seen in **Figure 4.17**, the closed-system pingos in Alaska and northwestern

CLOSED-SYSTEM PINGO

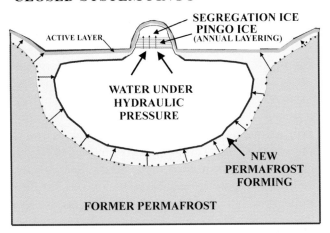

OPEN-SYSTEM PINGO

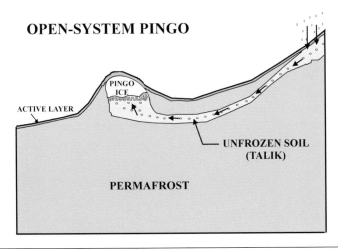

Figure 4.16 The mechanisms responsible for closed-system and open-system pingos.

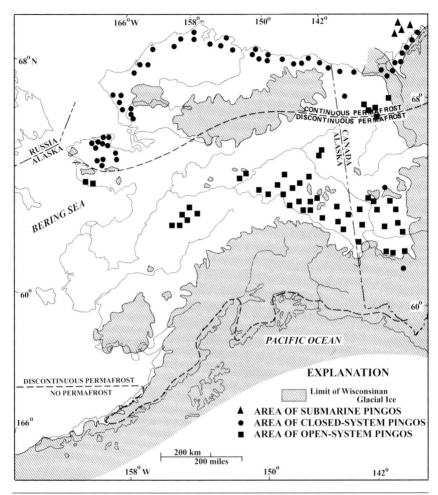

Figure 4.17 Distribution of open- and closed-system pingos in relation to permafrost zones and areas covered by Wisconsinan glacial ice in Alaska and western Canada. Modified from Figure 32 in Péwé (1975a).

Canada lie primarily in the low northern coastal plain. This is a region of continuous permafrost, where the mean annual temperature is –5°C or below. The open-system pingos are almost entirely in the warmer discontinuous permafrost region, and mostly in a zone of valley-cut upland within central Alaska that extends eastward into Canada. With a few exceptions, cold continuous permafrost is necessary if closed-system pingos are to form. Similarly, a warmer environment permitting discontinuous permafrost is required for the formation of open-system pingos.

Closed-system pingos

Closed-system pingos are formed on low ground primarily from water trapped when the boundaries of dying lakes freeze. Of nearly 1,400 closed-system pingos mapped in the Mackenzie Delta area of Canada, 98% were in or beside former lake beds.[13] The lakes once were deep enough to remain unfrozen, but vegetation growth, alteration of drainage, or other change allowed subsequent inward freezing from the sides and bottoms of the lakes as well as from the top. If a lake's freezing soil is sufficiently coarse-grained its volume remains essentially constant, but the freezing displaces water ahead of the freezing front. If the soil is fine-grained, the cryogenic suction $P_{ice} - P_{water}$ generated by the temperature falling below 0°C will draw water to and just behind the freezing front, thereby increasing the volume of the freezing material (frost heaving, in essence). In both situations, inward freezing creates a confining volume less than the volume of unfrozen water remaining so that this water is under increasing pressure. Part or all of the pressurized water may eventually freeze into what is termed *injection ice* or pingo ice, expanding by 9% as it does, to form a massive ice core that pushes the surface upward. However, the observed thickness of core ice formed in closed-system pingos and their overall height require pressures not attainable by hydrostatic pressure alone; thus at least a portion of the massive ice layers found in these pingos must be segregation ice formed from water pulled in by cryogenic suction. Since this suction depends on the grain size of soils as well as on subzero temperature, it follows that the fine-grained materials typical of many lake beds should foster the growth of large closed-system pingos.

As was observed by the Russian scientists who watched a pingo rupture, some of the water trapped within or below the pingo may escape by bursting out through cracks opened in the tops or sides of the pingo. Such loss of water probably is the explanation of the occasional observation of the periodic rise and fall of some pingo surfaces. Holes drilled by J. R. Mackay into one pingo on the Canadian Arctic Ocean coast released 30,000 to 40,000 cubic meters of water, and the top of the pingo sank by 60 cm.[14] Just as water may at times escape through dilitation cracking or unfrozen soil at the top of a closed-system pingo, surface water may at other times seep into a pingo from above. In fact, ice wedges have been observed on some pingos, although those wedges perhaps evolved before the pingos pushed upwards. **Figure 4.18** shows examples of closed-system pingos.

---

13. Stager (1956) cited by Washburn (1980) p 181.
14. Reported by Washburn (1980) p 185.

Figure 4.18A Two views of an egg-shaped ice core in a pingo cut away by wave action. Photographed by J. R. Mackay in 1955 on the Mackenzie Delta, near Point Atkinson, 100 km northeast of Tuktoyaktuk; see Figure 4.17. [One view is reproduced as Figure 79 in French and Heginbottom (1983) and the other is reproduced as Figure 4.11 in Pissart (1994).]

Figure 4.18B Layered ice in a pingo. J. R. Mackay who photographed the layers suggested that they are annual accumulations and that they have been tipped over since deposition. Reproduced from Figure 6.1b in Williams and Smith (1989).

Open-system pingos

Whereas closed-system pingos form on flat areas that once were lake beds, open-system pingos typically lie in valley bottoms or on their slopes. They tend to be smaller than the closed-system pingos. The artesian water supplying the open-system pingos enters the ground on hilltops or hillsides, and as it flows downslope (perhaps carrying along finegrained soil particles), it becomes trapped under permafrost so that it is under sufficient hydrostatic pressure to return to or near the surface through driven wells or natural channels. If the latter, the open water system may lead to the formation of one or more pingos, and they do tend to cluster. The interrelated association of pingos, permafrost and artesian well water is illustrated by an area a few miles north of Fairbanks, Alaska, where a road (Farmers Loop) runs generally east-west near the base of a southerly slope. Several wells adjacent to the road have yielded artesian water, and at least one pingo is nearby. Downslope from the road the permafrost layer is approximately 50 m thick, and part of the area contains extensive polygonal ice-wedge arrays. A few miles farther north, in a smaller valley, are a number of open-system pingos. Like the closed-system pingos, the open-system pingos often display split or depressed tops, but since they are mostly in areas of irregular topogra-

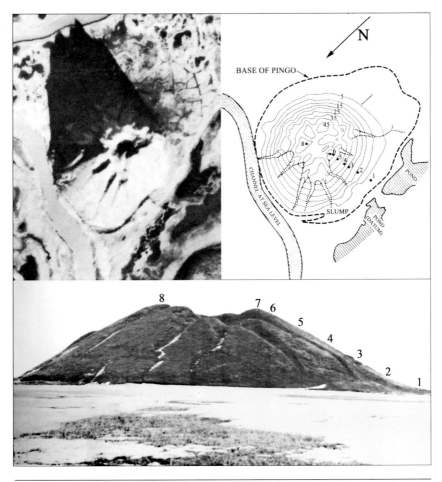

Figure 4.18C J. R. Mackay's ground photo and contour mapping of the Ibyuk pingo in the Mackenzie Delta near Tuktoyaktuk, along with a vertical aerial photo. Compiled from Figures 77 and 83 in French and Heginbottom (1983).

phy they do not stand out as dramatically from their surroundings as do the closed-system pingos. Shrubs and trees growing on the open-system pingos can hide them, but the vegetation on the mounds may differ enough from that nearby to attract attention, and thus help lead to their identification as pingos. Depressed tops or elongated depressions on the slopes of the mounds are identifying characteristics, but some might not be positively identified without drilling into their ice cores. **Figure 4.19** contains examples of open-system pingos.

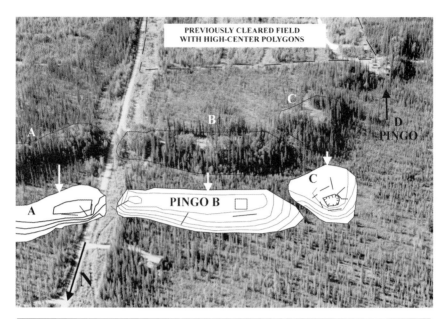

Figure 4.19 Open-system pingo complex on a 6-degree north slope in Goldstream Valley north of Fairbanks. On the offsets Labeled A, B, and C straight lines indicate ground cracks or wider collapse features. These also contain sketched-in contours at intervals of approximately 1 meter. The roadway (Miller Hill Extension) sloping downhill to the north cuts between pingos A and B, perhaps bisecting what was once a single pingo. A minor flow of water runs out of pingo B along the roadway. A house elevated on pilings sits atop pingo B, not far from where a previous structure had collapsed some years ago. The major collapse feature with numerous open fractures on pingo C resulted from clearing the vegetation several years ago. Houses sit atop the structure identified as pingo D. Uphill of the complex is an abandoned field now covered by brush and high-center polygons.

Pingo growth rates and life cycles

The available information on the ages of pingos and their growth rates comes mostly from observations of closed-system pingos. Many appear to be several thousand years old, but others are much younger, and some are so new that contemporary scientists have observed their birth and early growth. Carbon dating indicates that pingos at two widely separated localities on Banks Island in northern Canada grew during a cooling period occurring 4500 to 7000 years ago. It appears that many other such pingos are thousands of years old, but restricted to the Holocene epoch (the last 10,000 years).[15]

---

15. French (1977) cited by Washburn (1980) p 186.

Very young pingos also exist, and during the past 200 years observers have seen the initiation and early growth of a number of them. During the first year or so of pingo life the growth in height may exceed 50 cm per year, and the maximum reported is 150 cm per year. Based on his observations, the well-known Canadian geocryologist J. R. Mackay has suggested that, in early stages at least, pingo growth probably decreases with the square root of time. By 1972 he had seen pingos in the Mackenzie Delta that had formed after 1935 and that had grown to heights near 6 m, and also others birthed after 1950. Overall, the reports on growing closed-system pingos suggest rates typically near 10 cm per year and mostly within the range 0 to 50 cm per year.[16]

Inactive and fossil pingos (pingo scars)

When the entire volume of water available to build a closed-system pingo freezes or flows out through dilatation cracks, the pingo can no longer grow; even before that happens, dilatation cracking may expose the pingo's ice core to thawing that terminates the growth. Open-system pingos can suffer the same fate if the water supply is cut off or if dilatation cracking permits exposure to thawing. Climatic changes or lesser events such as forest fires that modify the vegetation can also stop pingo growth and foster eventual collapse.

Fossil pingos—the remains of formerly active pingos—may appear as pits not easily distinguished from other kinds of thermokarst depressions, but the pits often are surrounded by rampart-like rims that point to their pingo origin, as shown in **Figure 4.20**. Some fossil open-system pingos might long retain at least a part of their mound-like character owing to deposition in the pingo structure of fine-grained soil particles carried in with water through the supply conduits. At least one of the pingos shown in **Figure 4.21A** contains considerable clay in the upper meter or so of ice, as shown in **Figure 4.21B**, suggesting that these are open-system pingos. Some fossil open-system pingos are in the form of elongated, sometimes U-shaped rampart-like features that open uphill so as to enclose marshy areas. Some cross over each other, suggesting repeated episodes of pingo growth during favorable conditions.

Wherever fossil pingos can be identified in nonpermafrost areas they serve as proof of former cold climate. Extensive groups of fossil pingos of Pleistocene age have been found in the British Isles, Europe, and Asia. As many as 35 Pleistocene fossil pingos per km² have been identified in one

---

16. Based on discussion by Washburn (1980), pp 180–87.

Figure 4.20 Collapsed open-system pingo in Goldstream Valley, just north of Fairbanks, Alaska. The pingo collapsed some years ago after being cleared of trees in an attempt to prepare the area for agriculture: A) Close-up view of one of the many mud cones formed by mud-charged springs in the lake of the collapsed pingo, the one labeled A in Part C; B) Aerial view of the collapsed pingo; C) View across the pingo lake with mud cones marked by arrows. These indicated water coming into the feature through numerous channels. New collapse around the perimeter in 1996 caused the dip in the road indicated by the tipped vehicle. See also Plate 10, bottom, showing two mud cones.

locality of northern France, but a density of one pingo per square kilometer is more typical.[17] The oldest reported fossil pingos are ones that developed in the Sahara during Ordovician times (500 million years ago).[18]

---

17. Cailleux (1976) cited by Washburn (1980) p 190.
18. Beuf, Bernard et al. (1971) cited by Washburn (1980) p 189.

Figure 4.21A Described both as open-system pingos (Péwé, 1977) and palsas (Péwé, 1997), these large mounds stand adjacent to the Maclaren River in the Alaska Range at Mile 40.8 on the Denali Highway. An array of sorted polygons lies in the foreground. Photographed from the hill just to the east in 1996. The peat in contact with the ice in one mound is approximately 10,500 years old (Péwé, 1977).

Pingo or Palsa?
Dirt inclusions in the upper portion
of the ice (see below) imply water
movement under pressure,
the mechanism for forming pingos.

Figure 4.21B Beneath the 3-m-thick covering of peat, ice within the mound at the right-hand edge of Figure 4.21A, shows extensive thin layers of clay in the top meter of the ice. The amount of clay decreases with depth so that below depth 1 meter in the exposed cut, the ice contains only occasional thin lenses of clay. Photographed in August 1997.

Palsas

Like pingos, *palsas* are ice-cored mounds that bulge up above their surroundings. They tend to be smaller than pingos and more varied in shape. Palsas are from 1 to 12 m in height and range in shape from conical through hump-like and plateau-like to elongated ridges or esker-like dikes. The main difference between pingos and palsas is that palsas grow from water lifted by cryogenic suction instead of that transported by the pingo-forming combination of cryogenic suction and hydrostatic pressure.[19] Dilatation (doming), frost contraction or desiccation cracks typically develop on palsa surfaces and lead to their eventual demise.

Palsas form in wet boggy areas where a combination of mean annual temperature less than −1° or −2°C and low snowfall or high winds expose any projection above the ground surface to the winter cold. The result is a higher temperature gradient in the projection than in the surrounding

---

19. Seppälä (1988).

mossy area and thus a condition there that favors the formation of layers of segregation ice that frost-heaves the surface even higher. Contributing to the ongoing growth of palsas over the course of many years is the seasonal change in thermal conductivity of the covering of peat-like material (refer back to Table 2.1). Frozen peat is about five times as conductive as unfrozen water-saturated peat and nearly thirty times more conductive than dry peat.[20] Consequently, heat flows out far more rapidly through the top of a palsa in winter than it flows in during dry summer conditions.

When arrays of ice wedges become inactive and melt enough to form high-center polygons, the elevated center parts of the polygons may sometimes become the bases from which palsas grow. Other palsas emerge from the boggy areas typically formed from vegetation growth in shallow lakes and former lake areas. Vertical growth rates up to 15 cm per year have been observed, but average rates probably are somewhat lower. Observations of palsas in Finland indicate that their ages range from 100 to 2000 years,[21] and dating of the moss in a palsa complex in central Alaska (shown in **Figure 4.22**) indicates a possible maximum age of about 10,000 years.[22] Since the cores of most palsas are primarily ice—reportedly as much as 80 to 90%— palsas are like pingos in that they contain the seeds of their own destruction. The palsas grow until dilatation or other cracking, or perhaps disturbance from wildfire, breaks the surface enough to create melting and collapse, and then new palsas may grow at the same general location. Because of this cyclic growth behavior, palsas of various ages are found together. When they do collapse, palsas may leave little evidence of their former existence, but some leave minor rampart-like rims that may be quite difficult to distinguish from those of small fossil pingos. And even before they collapse, palsas may exhibit characteristics so similar to those of some open-system pingos that it is not easy to tell them apart.

Hummocks.   Hummocks are small mounds to 50 cm diameter and to 1 m in height that form in uniform arrays on level or near-level ground in both permafrost and non-permafrost areas. They are of two main types: those called *earth hummocks* that contain cores of mineral soil, and those called *turf hummocks* that contain mainly organic material.[23]

---

20. Seppälä (1995).
21. Ibid.
22. Reger (1995).
23. The discussion here is based on material presented by Washburn (1980) pp 121–47; Williams and Smith (1989) pp 157–63; and Lundqvist (1969).

Figure 4.22A Palsas in an array at Mile 206.7 on the Parks Highway just south of Cantwell, Alaska. Dating of basal moss on the palsa gives an age of 9200 years (Richard Reger, private communication 1995).

Figure 4.22B Close view of nearly pure ice exposed within the palsa in foreground of Part A.

Figure 4.22C A symmetrical palsa mound in the complex; see also Plate 11.

Figure 4.22D An elongated palsa in the background with one that is collapsing in the foreground; see also Plate 12. Photographed fall 1995.

Earth hummocks (including *thufur*, a form identified in Iceland and England) develop in fine-grained soils prone to frost heaving in environments where vegetation cover is light to moderate. The hummocks typically are dome shaped or sometimes elongate if on gentle slopes. They can occur singly but usually are in groups with spacings generally less than the diameter of the hummocks. That they can develop rather rapidly in non-permafrost areas is indicated by reports of their forming within one or two decades in English and Icelandic pastures. Earth hummocks in permafrost areas indicate actively heaving surfaces (as shown sometimes by tilting trees growing on the hummocks) and those hummocks are associated with accumulations of segregation ice below the permafrost table. These hummocks may be thousands of years old; dating indicates that the period 2500 to 5000 years ago was a good time for hummock formation.[24]

Seasonal frost, adequate moisture, and heave-prone soil are necessary if earth hummocks are to form. What starts the hummock forming process is not clear, but recently reported theoretical work suggests that spatial differences in the amount of frost heave can be predicted and that the spacing between hummocks is likely related to the depth of frost penetration at the time when certain circumstances involving the rate of frost penetration and other factors are such that they promote rapid heaving.[25] This prediction is in keeping with observations that even artificially leveled soil that is fine-grained enough to be prone to frost heaving tends to develop an irregular surface. Once the surface is irregular, differential rates of freezing and accumulation of segregation ice—aided by differences in vegetation in the high and low areas—promote hummock growth. The tops of the embryonic hummocks are likely to be drier than the surrounding troughs during freezing, and the tops tend to develop a more insulating vegetative cover. Thus, initially, freezing should penetrate most rapidly in the troughs and the accumulation of pore and segregation ice there should squeeze the more slowly freezing soil beneath the hummock upward, thereby enlarging the mound. Later, at least in permafrost areas, a thicker vegetation layer grows in the troughs and then the active layer becomes thickest at the centers of the earth hummocks. There the greater depth of freeze and thaw each year creates bowls in the permafrost that may collect water during thaw periods. If thawing is rapid, high enough pressure may develop to help lift soil upward in the hummocks. However it occurs, the motion evidently is turbulent because

---

24. Zoltai et al. (1978).
25. Fowler and Noon (1997).

Figure 4.23 Top: A turf (bog) hummock; bottom: an earth hummock showing cryoturbated soil layers, both photographed along the Denali Highway between Miles 30 and 37 in 1997. See also Plate 13.

the hummocks often show irregular streaking of soils, and the carbon-dated ages of adjacent layers can vary wildly, by more than 1000 years.[26] This churning (cryoturbation, a process discussed more below) extends down only to depths of 30 to 60 cm in earth hummocks of nonpermafrost areas, but the churning reaches to the bottom of hummocks nested in permafrost. **Figure 4.23** shows an earth hummock, and also a turf hummock.

Turf hummocks (also called bog and peat hummocks) are of similar size or smaller than earth hummocks and often more columnar, but like them can develop in both permafrost and nonpermafrost settings. They are

26. Zoltai et al. (1978).

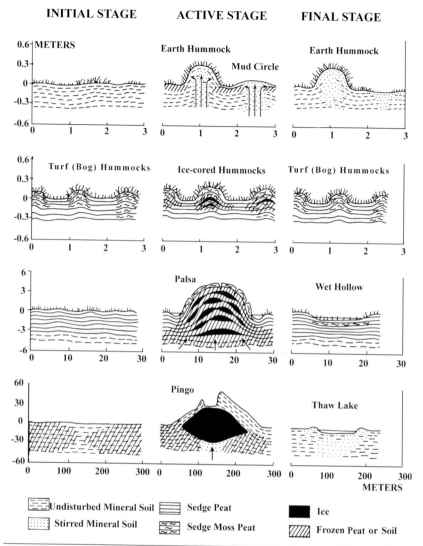

Figure 4.24 A schematic diagram depicting the differences among pingos, palsas, earth hummocks and mud circles. Modified slightly from Figure 1 in Lundqvist (1969).

widespread in boreal forest and tundra areas where they make walking difficult. The turf hummocks consist almost entirely of sedge (such as cotton grass) and sphagnum moss (also called peat moss), and are the consequence of differential vegetative growth rates rather than cryogenic transport of material. Some may at times contain cores or layers of segregation ice but when that ice melts the hummocks retain their upright form.

Figure 4.24 provides a useful summary of the general characteristics of pingos, palsas, and hummocks. It is sometimes difficult to distinguish among these various forms of *frost mound* and similar forms created by other mechanisms; part of the problem can be that transitional forms develop.[27] Hydrostatic pressure causes pingos, cryogenic suction generates palsas, and frost heaving is responsible for earth hummocks, but all of these mechanisms might operate simultaneously in certain circumstances, including the differential rates of vegetative growth that form turf hummocks.[28]

▪ Weathering, Sorting, and Transport Processes in Cold Lands

Although many cryogenically produced landforms—pingos, palsas, polygons, and the like—occur in mossy or otherwise heavily vegetated lowlands, others are located in high alpine or arctic upland areas where trees are mostly absent and the ground vegetation is far from lush. There, cryogenic weathering and transport processes play a major role in shaping the land, and most important are those related to cyclic variations in temperature, mainly in the active layer. Less effective because of low temperature is weathering due to chemical reactions.

Chemical weathering

In warm, moist climates the most important cause of weathering is chemical. Chemical weathering involves several specific mechanisms that alter the rocks and cause their decomposition: 1) Dissolution, wherein water directly dissolves rocks or combines with other molecules to form rock-dissolving acids or other substances; 2) Oxidation, wherein oxygen combines with mineral ions contained in rocks to form insoluble but generally weak oxides; 3) Acidification, wherein acids derived from decomposing organic material decompose soil minerals; and 4) Hydrolysis, wherein water molecules combine with minerals and the H+ or OH- ions contained in water replace other ions in mineral structures to create fine-grained materials such as clay and also soluble compounds readily transported away by water. Heat and moisture foster chemical weathering, and living organisms assist by helping to expose rocks to weathering (specific processes include root growth by plants and soil churning by earthworms and other animal life). Granites, cherts, gneiss, and quartzite rocks resist chemical weathering; sandstones, basalts,

---

27. An example of the problem is the mound complex shown in Figure 4.21 which has been described both as open-system pingos [Péwé (1977)] and palsas [Péwé (1997)].
28. Nelson et al. (1992).

and schists are intermediate; and limestones and shales are least resistant. Chemical weathering tends to produce rounded rocks and landforms.

Cryogenic weathering

Cryogenic weathering, like chemical weathering, proceeds by several identifiable but interrelated processes, but each is of strictly mechanical nature. Cryogenic weathering pries rocks apart and mechanically chews them up into little pieces rather than chemically altering them, and it proceeds most rapidly where an ample supply of water is present and the temperature cycles frequently through the freezing point. Cryogenic weathering is commonly called *frost wedging* (and occasionally gelifraction, congelifraction, or gelivation, all words derived from the Latin *gelare*, meaning to freeze). The specific mechanisms are:

1. The freezing of water seeping into preexisting structural cracks, wedging rocks apart to create smaller fragments.
2. The freezing of water seeping into rocks along grain boundaries or other zones of weakness to form ice layers that pry splinters and thin sheets from the rock surfaces.
3. The freezing of water contained in the pore spaces of porous rock bodies, shattering their outer parts into fine-grained particles.
4. The damaging expansion and contraction of rocks caused by radical changes in the amount of adsorbed water on pore surfaces as the temperature varies. Called *hydration shattering*, this mechanism can operate at temperatures above freezing, but because temperature fluctuations into the subzero range so radically affect the free energy of water in pore spaces, thereby affecting how many molecules get adsorbed onto the pore surfaces, hydration shattering may be particularly potent in this subzero range. Its effectiveness relative to the freezing of water in pore spaces is uncertain, perhaps since both mechanisms work to give the same result: the shattering of a rock body into small particles. In principle, hydration shattering should most affect fine-grained rocks such as shales and siltstones because of their high surface areas.[29]

---

29. In his excellent textbook *Geology* (1995) Stanley Chernicoff cites an experiment in which highly polished granite showed no effect when cycled through a temperature change of 38°C 89,500 times (equal to 244 years of daily cycle) but when the experiment was repeated using a fine spray of water to simulate nightly dews the surface of the granite became irregular and cracked.

5. The prying apart of rock by the formation of segregation ice from water transported into porous rock by cryogenic suction to produce splinters and rock fragments. Enormous pressures—up to 300 Atm—can develop according to the premelting model of segregation ice formation.

The susceptibility of rock to cryogenic weathering depends on a number of factors including the internal structure of the rock and its porosity, the availability of water, and the range, frequency, and speed of temperature fluctuations. Porous rock containing water typically grows stronger when the water freezes, and its strength increases with decreasing temperature. However, the internal pressure generated by the freezing of ice in rock pore spaces also increases with decreasing temperature, theoretically to a value roughly 10 times the tensile strength of rocks like sandstone and limestone (which have tensile strengths near 250 Atm). Thus, how cold it gets during a cycle of temperature fluctuation can help determine how much weathering occurs during that cycle. Various laboratory experiments indicate that the breakdown of rocks depends on the intensity and speed of freezing and also on the number of freeze-thaw cycles undergone. For example, when testing porous rocks, one experimenter found little cracking until the temperature was below −10°C, and the cracking occurred only when the rate of temperature decline was at least 6°C/hr.[30] Cryogenic weathering produces angular fragments ranging from house-size blocks down to those the size of sand, or even clay in certain circumstances.[31] Open-air settings (such as rock cliffs) and environments of extreme cold tend to produce a high proportion of large angular block debris, whereas situations involving rocks buried in debris (so that temperature fluctuations are not as rapid or severe) lead to a higher proportion of fines.[32]

Laboratory experiments in general also demonstrate that the availability of water is very important: rocks partially or fully immersed in water break up more rapidly than those coated only by films of water. Field observations show that rocks near patches of thawing snow or near shores are particularly prone to disintegration and fracturing when repeatedly frozen and thawed.[33]

---

30. Battle (1960) cited by Washburn (1980) p 75.
31. McDowall (1960) cited by Washburn (1980) p 76.
32. Tricart (1970) p 74.
33. Tricart (1970) pp 73–76.

The nature of the rock is a major factor as well. Crystalline (igneous) rocks such as granites tend to be the least permeable unless they contain the sheet-like muscovite and biotite micas that provide avenues along that water can enter and then wedge the rocks apart. Particularly susceptible to cryogenic weathering are fine-grained sedimentary rocks such as shale and siltstone that have a water-admitting layered structure or which contain plate-like clay or mica minerals. Individual minerals contained in rocks react differently to cryogenic weathering: quartz (SiO_2) is particularly susceptible, especially if heavily fissured, and biotite mica is more susceptible than muscovite mica.[34]

Those cryogenic weathering mechanisms that operate on a micro-scale—as contrasted with the large-scale prying apart of rocks into angular blocks—tend to produce particles of rather small size. They can split up chemically weathered clays into even smaller particles, but the end products of the weathering also depend much on the nature of the parent rock. Chemical weathering of quartz yields smaller particles than chemical weathering of feldspars, but cryogenic weathering of quartz produces larger particles than does cryogenic weather of feldspars. This difference seems to be a consequence of the repetition of freeze-thaw cycles, and this repetition is highly effective in splitting the feldspar-derived clay particles. The repeated freeze and thaw cycle also modifies soil, altering both soil particles and the size and shapes of pore spaces. Soil particles near layers of segregation ice tend to become more closely packed, and when the ice thaws the soil particles do not necessarily return to their prefrozen locations.[35]

Churning and sorting within the active layer

Cryogenic weathering typically leads to a soil layer containing products of various sizes, and if the size distribution and water availability are such that the soil heaves upon freezing, then a sorting of the material occurs quite naturally. **Figure 4.25** schematically illustrates the process, a shifting of objects that depends directly on their dimension in the direction of freezing, assuming that the freezing soil will grab tight to the object somewhere above its base and thereafter lift it as frost heave continues. Geocryologists generally refer to this as the *frost pull* mechanism. By the same means, any object such as a rock that has one end elevated above the other will tilt upward even more as freezing takes place. Since various

---

34. The discussion here is based primarily on material given by Lautridou (1988).
35. Williams (1999).

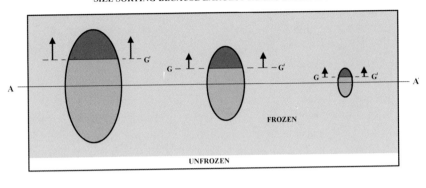

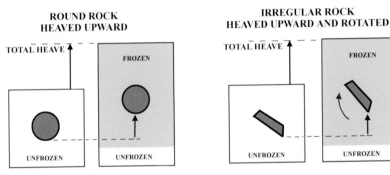

Figure 4.25 Schematic illustration of: top; how if similarly shaped rocks of various sizes lie at the same level in a soil profile, the frost pull mechanism of frost heaving can sort them by shifting them upward in proportion to their sizes because the soil that is heaved the most (the upper part of the profile) grasps the larger rocks, leaving the smaller ones to be lifted later by lesser amounts as the frost line penetrates the profile, and at bottom; that round stones merely heave upward (or perpendicular to the plane of the freezing front) but that heaving soil tilts irregular rocks by grasping their upper parts first, assuming the unfrozen soil below allows them to rotate.

observations show that frost heave does cause stones, fence posts, and other objects to rise out of the ground over the course of years, it is evident that at least part of the uplift occurring during freezing is preserved during subsequent annual thaws. Thus, vertically oriented rocks protruding from the ground are a common sight in certain cold land areas (see **Figure 4.26**). Dirt trickles down into the voids left below frost-heaved objects, and sometimes the friction between the objects and adjacent unfrozen material is sufficient to keep them elevated. Sometimes people

Figure 4.26 Uplifted rocks protrude through the thin moss cover of Eagle Summit, north of Fairbanks, Alaska, creating a moon-like landscape that provides a good habitat for rodents, which in turn titillate other animals like the 1-meter scale dog (nose to tail) in the right-hand picture.

have been able to hammer back into the ground posts elevated by frost heave, indicating that voids remained below.

As the freezing front progresses down across a rock or boulder embedded in soil, it may move faster through the rock than through the adjacent material simply because the rock has no latent heat of fusion to release, and perhaps also because the rock has relatively high thermal conductivity. The consequence then may be an accumulation of segregation ice below the rock that helps lift it. The overall effectiveness of this *frost push* mechanism is uncertain. Some of the water forming segregation ice beneath the rock may actually be migrating downward from wet soil beside the rock. If that soil is loose, some of it may be drawn downward with the water, thereby amplifying the sorting process, but another complicating effect may be desiccation and consequent volume decrease of the soil near the rock. Frost push appears to be important in lifting blocks of bedrock if that rock is porous, and if thermal conditions are such that the coldest part of the rock is well below the surface, perhaps resting on cold permafrost so that the freezing front is advancing upward within the block. Water can then move downward through the porous stone to form segregation ice that lifts the bedrock through the unfrozen soil. Experiments and field observations have shown that this mechanism can lift porous bedrock by approximately 5 cm annually.[36] In essence, the porous rock is acting like a fine-grained soil in permitting the accumulation of segregation ice.

Cryogenic sorting of soil and the upward displacement of rocks it contains can proceed even if the soil is generally texturally uniform throughout. If the soil is nonuniform, which is usually the case, differences in heat capacity and thermal and hydraulic conductivity are likely to lead to even more complex motions of water molecules, soil particles and contained rock fragments. The overall assemblage of churning and vertical uplift motions thereby taking place in the active layer goes by the name cryoturbation and, as the turbulent part of that name implies, the motions can be complicated. One part of the soil profile might expand by accumulating water and segregation ice while another contracts from loss of water or thawing. Further complication comes from the possibility of upward freezing from a permafrost layer as well as downward from the ground surface. The pressure within unfrozen pockets of wet fine-grained material caught between growing frozen layers may cause the material to squirt to a different place or break through to the ground surface. If the

---

36. Price (1970) [cited by Washburn (1980) p 91].

water contained in saturated fines cannot readily escape as they thaw, excess pore pressure can build up and the material may become so unstable that it cannot support the overlying load, causing slope failure or flow. Evidence of past cryoturbation includes deformed soil profiles ranging from wavy through regularly folded to highly complex, sometimes chaotic contortions, all of which fall under the generic name *cryoturbations*, also *involutions*.[37, 38] **Figure 4.27** contains schematic sketches showing some types of cryoturbations and also drawings of observed examples.

Cryogenic transport of weathered material

Water and wind operate worldwide to erode and transport weathered rock material, and in the process usually achieve some sorting. Wind, for example, carries fine particles farther than large ones from a source area, and water tends to do likewise. Both of these transport mechanisms abrade the moved material and in doing so round it into smoother forms. Deposits of rounded gravels and stones are sure signs of transport by fast-moving waters. By contrast, cryogenic transport processes tend to leave the material in angular shapes. Owing to low precipitation in most of the arctic and subarctic region, erosion by water is slow and episodic, so freeze and thaw processes are responsible for much of the transport of weathered material. The material thus transported remains angular, and the transport mechanisms also create landforms unique to cold lands. However, those landforms are influenced by eroding waters that flow across and beneath sloping surfaces (*slopewash*). Surface runoff from snowmelt and summer rains can transport fines carried in suspension and also remove soil by carrying it in solution. Subsurface waters can carry away dissolved soil, and also that held in suspension when the subsurface water is flowing down through voids in rocky deposits, a process called *piping* or *stoping*.[39] The few measurements of rates of erosion by slopewash in cold areas indicate very slow transport that may lower surfaces by only a few millimeters in 1000 years.[40]

Cryogenic transport—the downslope movement of soil that is frozen or undergoing freeze and thaw—proceeds by three general processes: 1) very slow creep within the body of perennially frozen soil (*permafrost*

37. The term "periglacial involutions" also is used, and some geocryologists consider involutions to be a special form of cryoturbations [Washburn (1980)] p 170.
38. Vandenberghe (1988).
39. Péwé and Reger (1983) pp 125–27; Lewkowicz (1988) pp 354–57.
40. Lewkowicz (1988) pp 354–55.

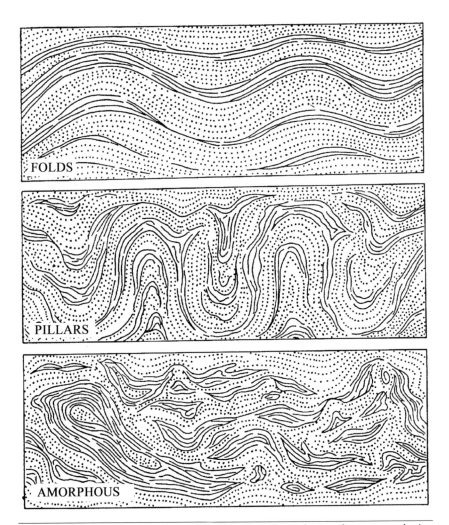

Figure 4.27A Schematic drawings of folded, pillar-like and amorphous cryoturbation structures.

creep), 2) slow *solifluction* flow, and 3) relatively rapid flows associated with slope failures of one sort or another. Solifluction proceeds by three different identifiable mechanisms: 1) *frost creep*, 2) *needle ice (pipkrake) creep*, and 3) *gelifluction*. The rapid cryogenically caused mass movements may involve viscous-like flowing of unfrozen, often-saturated soil material, rapid en masse sliding along slippage planes, and falling away of blocks detached by freezing and thawing.

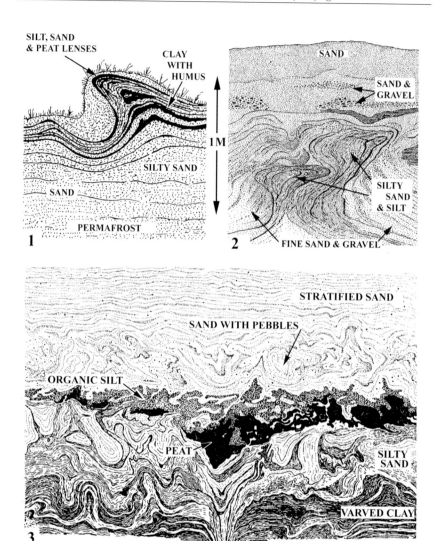

Figure 4.27B Drawings of actual cryoturbation structures observed within Pleistocene deposits in Poland: 1) a fold resulting from differential frost heaving; 2) pillar structures; 3) complex folded and pillar structures. Compiled from Figures 89, 92 and 93 of Jahn (1975).

Permafrost Creep

Given enough time, frozen ground, even if under light loading, will undergo creep. In warm permafrost, that at temperatures just below 0°C, enough unfrozen water may be present to enhance creep by reducing the friction between soil particles, especially if the permafrost is ice rich. Observations in the Mackenzie River area over a period of several years on a steep slope (15 to 30°) containing both dry sand and ice-rich clay showed little creep in the sand and more in the clay, where it appeared that deformations were occurring in conjunction with ice layers. The creep was slow, 2.5 to 3.0 mm/yr, but fast enough to indicate that creep within warm permafrost may be an important process in helping to denude steep slopes of soil.[41]

Solifluction

About a century ago, the term solifluction was used as a name for slow downslope flow of soil saturated with water. Over the years the meaning has evolved to include slow downslope flow caused by or associated with cryogenic processes.[42] In the interest of fostering clarity, I here use the term *cryogenic solifluction* specifically to mean solifluction caused by or associated with cryogenic processes. The term cryogenic solifluction (or simply solifluction, if the meaning is clear) is useful for refering to the full gamut of specific physical mechanisms involved: frost creep, needle ice creep, and gelifluction.

Frost (active layer ratcheting) creep. Freezing of a sloping soil in the active layer causes it to push upward in the direction perpendicular to the slope. Then, as it thaws, the material tends to drop straight down. Thus, repeated freeze and thaw rachets soil downslope, as illustrated in **Figure 4.28**. Observations show that frozen material does not necessarily drop vertically, but instead falls back slightly toward its original position, thus executing a retrograde motion. That behavior is thought to be due to a cohesion between the soil particles that fosters maintenance of the original positioning of one soil particle above another, or perhaps a relative desiccation of the top part of the profile that causes it to contract to slightly smaller volume than the soil below.[43] The few measurements made on frost creep in the active layer indicate that it progresses about ten times faster than creep in permafrost, at rates up to approximately 2 cm/yr.

---

41. Much of the discussion here is based on Lewkowicz (1988).
42. Washburn (1980) pp 198–213; Lewkowicz (1988) pp 329–39; Williams and Smith (1989) pp 123–31.
43. Washburn (1980) p 199. Also, this cohesion and the resulting cracking may well be a major factor in determining the spacing of cryoturbation steps.

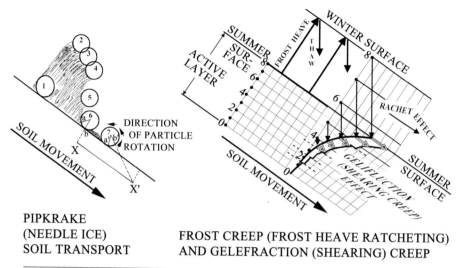

PIPKRAKE
(NEEDLE ICE)
SOIL TRANSPORT

FROST CREEP (FROST HEAVE RATCHETING)
AND GELEFRACTION (SHEARING) CREEP

Figure 4.28 Processes causing solifluction. Left: downhill transport of soil by pipkrakes (needle ice) that also causes retrograde rotation of soil particles [reproduced with minor modification from Figure 95 of Jahns (1975)]. Right: Both frost creep (frost heave ratcheting) and gelifluction shearing in the active layer may act to cause the upper part of the active layer to override the lower part. Dashed lines suggest locations of segregation ice layers. In the diagram frost heaving and shearing are both exaggerated for purpose of illustration; not shown is the slight retrograde motion of soil involved in frost heave ratcheting.

Needle ice (pipkrake) creep. In climates where repeated freeze and thaw occurs and where soil permeability and adequate moisture combine to allow the growth of tall pipkrake crystals, they can be an important and rapidly acting agent for moving the very top of the soil layer. Needle ice creep moves surface material downslope at rates typically near 10 to 20 cm/yr, and in some instances more than 30 cm/yr. In cold-climate areas, needle ice creep probably is of less importance than the other cryogenic solifluction processes: frost creep and gelifluction.

Gelifluction (shearing creep). Gelifluction is cryogenic solifluction involving shearing within the active layer because of slow slippage, probably mainly along melting layers of annually formed segregation ice in consequence of particularly high pore pressures developed there. Within a thin layer at the thawing interface of an ice-rich soil, the pore pressure can become so high that the soil's strength is drastically reduced until downslope stresses exceed the soil's resistance to shear. Gelifluction can develop on slopes as low as 1 or 2°; silty soils are particularly susceptible.[44] Contributing to gelifluction may be a variety of microscale differential motions related to expanding and contracting soils, reorienting of soil particles and to cavities formed in the soil by air coming out of solution during freezing. Some of these may progress during the freezing process or after the soil is frozen, so gelifluction is a process that can be underway much of the year.

Especially in situations where the active layer is in contact with continuous permafrost, the heaving associated with frost creep, like gelifluction, can occur in summer as well as winter, and then the distinction between the two forms of solifluction blurs so much that it perhaps becomes artificial.[45] Both kinds of solifluction typically lead to vertical velocity profiles that are concave downslope (as in Figure 4.28), so that the combination of the two kinds of solifluction causes material to override itself. The overall end result, notes Alaska geocryologist Richard Reger, is that the soliflucting material behaves much like the tread of a caterpillar tractor in that it rolls forward downhill over itself. In **Figure 4.29**, he points to some of the evidence for that suggestion.

Complexities in the gelifluction slippage along layers of melting segregation ice can modify the normal vertical velocity profile, causing it to be more irregular than illustrated in Figure 4.28, and even convert its concave-downslope shape to convex-downslope. This situation is most likely to occur when the active layer is in contact with permafrost, because then layers of segregation ice may form near the bottom of the active layer, especially in late summer.[46] The melting of these lower ice layers creates a shear zone that allows the overlying material to flow more or less as a unit, a process called *plug flow*,[47] which perhaps should be included with the other cryogenic solifluction processes.[48]

Cryogenically caused rapid mass movement

Characterized by their episodic and spatially discontinuous nature, cryogenically caused mass movements are important transporters of soil, especially on steep slopes, those exceeding 15 or so degrees. Viscous-like flowing of soil material and en masse sliding along slippage planes can

---

44. Washburn (1980) pp 202–04.
45. Lewkowicz (1988) p 332.
46. Mackay (1981).
47. Rein and Burrous (1980).
48. Williams and Smith (1989) p 130.

Figure 4.29 Geologists Richard Reger and Stu Rawlinson discuss the effects of solifluction, in this instance a deformed soil layer. Photographed 1978 in a roadcut on the campus
of the University of Alaska Fairbanks.

occur simultaneously, although in any particular situation one process may
dominate. The term *active-layer failure* has been used in a general way to
refer to any type of failure occurring strictly within the active layer overlying permafrost, and also as a name for flow-dominated mass movements
on steep slopes. Synonymous names for flow-dominated movements
within the active layer are *skin flow* and *earth flow.* These failures, typically
ribbon-like, with a small slump scar compared to the movement track,
may be only a few to several tens of meters wide and hundreds of meters
long.[49] The one shown in **Figure 4.30** exceeds 500 m in length. Causes of
the episodic flows include unusually rapid snowmelt or unusually heavy

49. Lewkowicz (1988) pp 342–42.

Figure 4.30 A ribbon-like active-layer failure (skin flow) originating high on a steep slope and flowing down a gentler slope almost to the Dalton Highway in Alaska's Brooks Range. Photographed in summer 1995.

summer rainfall that increases stress through saturation of organic material, **thaw consolidation**, and melting of permafrost caused by natural or man-made disturbance to ground cover or climatic warming.[50]

Another widespread active-slope failure on steep (>10°) slopes in permafrost areas is **active-layer glide** (also sometimes called **block slide**, **detachment failure**, and active-layer detachment). It involves downhill sliding of the active layer as a unit, usually along the typically ice-rich per-

50. Lewkowicz (1988) p 342.

mafrost table. A large area of the active layer may slide as a unit, but most commonly a section near the base of the slope detaches and then, over the course of minutes to days, sections higher up successively detach to create an overall slide area usually with length greater than width. The cause of active-layer glide is excess pore-water pressure near the bottom of the active layer. Heavy summer rainfall can contribute by increasing the weight (and hence the shear stress), and other contributors are man's activities or forest fires that deepen the active layer enough to melt the ice-rich soil just beneath the former permafrost table.[51]

Another flow-dominated episodic transport phenomenon is the *mudflow*, the downslope movement of saturated soil material. Mudflows are not necessarily created by cryogenic processes nor restricted to the active layer, so they exist wherever conditions foster saturation of soils on slopes. Cryogenic mudflows are usually small—only a few meters in extent—and so do not constitute an important transport process. In permafrost areas mudflows usually develop in consequence of thawing ground ice that has become exposed for one reason or another. They have been observed to emanate from mudboils, along eroding riverbanks, and on slopes with subsurface water derived from melting snow or summer rains.[52]

Still another flow-dominated failure of slopes in the discontinuous permafrost zone is *multiple retrogressive flow*, wherein repeated slope failures occur in a way that exhibits flow (typically within silty clay) but which in the bowl of the flow creates arcuate ridges retaining some evidence of the former relief.[53]

Differing from the multiple retrogressive flow is the *multiple retrogressive slide*, a large-scale repeated slumping that may develop in frozen and unfrozen sediments and along the permafrost table. The distinguishing characteristic is the production of a series of arcuate slumping blocks, concave downward and up to 60 m high, that step up toward the head of a scarp and tend to tilt over backwards, especially near the toe of the slide area.[54]

Another shear failure is *thaw slump* wherein blocks of frozen or unfrozen sediments break away along a concave slip face. Thaw slumps are common near rivers with undercut banks, particularly where sand or gravel overlies silt and clay. Contributory may be the development of high pore-water pressures in the fine-grained soils when the river banks freeze.[55]

51. Lewkowicz (1988) pp 344–45.
52. Lewkowicz (1988) pp 342–45.
53. McRoberts and Morgenstern (1974a; 1974b).
54. Lewkowicz (1988) p 345.
55. Lewkowicz (1988) p 345.

Rockfall, the free falling of rocks down steep slopes, is primarily a vertical transport process that operates in all climates, but it is a periglacial process in localities where freeze-thaw cycling is an important release process. Frost wedging, perhaps augmented by thawing of interstitial ice and stress increases created by avalanches or debris loading, is a significant cause of rockfall in cold-climate areas. The variation in frequency of periglacial rockfalls with season tends to follow variation in the number of freeze-thaw cycles; for example, rockfalls on south-facing slopes tend to occur earlier in spring than on north-facing slopes, and rockfalls tend to occur more frequently in spring and fall than in summer and winter. The rockfalls also tend to occur most frequently during the warmest part of the day. Although amounting to only a few millimeters per year at most, the combination of frost wedging and rockfall on near-vertical slopes results in substantial erosion—up to several tens of meters during the past 10,000 years (Holocene).[56]

▪ Landforms Resulting from Weathering, Sorting, and Transport Processes

Talus slopes

Frost wedging of steep cliffs contributes to the development of apronlike deposits (also called scree) below with their surfaces lying at the angle of repose. The tendency for the largest falling rocks to roll leads to some sorting. Any type of weathering can create talus slopes, so they are not limited to cold environments. Long-lasting snowbanks accumulating at the bases of talus slopes sometimes provide platforms for material to scoot across. The resulting deposits, called ***protalus ramparts,*** are thus isolated from the main talus slopes and may show up as separate ridgelike deposits best seen after climatic change reduces the accumulation of the perennial snowbanks.

Solifluction sheets and lobes

Cryogenic solifluction taking place on shallow (down to 1°) and steeper slopes creates various shapes of deposits: sheet-like, bench-like, and lobate. A characteristic of solifluction is that it tends to fill in low places, and so acts as a great leveler. Water erosion, by contrast, tends to deepen depressions and thereby acts to generate relief. Examples of the difference are widespread in the upland and mountainous areas of central Alaska where past or current cryogenic processes have created smooth, low relief slopes at high altitude,

---

56. Washburn (1980) pp 76–77; Lewkowicz (1988) pp 345–47.

Figure 4.31A Top: Solifluction lobes seen when looking west from Eagle Summit on the Steese Highway north of Fairbanks, Alaska. Bottom: Others seen a few miles south of the summit.

and water erosion has incised the slopes lying at lower altitudes where past and current climate favors water transport over cryogenic transport.[57]

Among the more easily recognized sheet-like landforms created by solifluction are the turf-covered *solifluction benches* and lobes that spread out across upland slopes in north-central Alaska and elsewhere, the benches tending to develop more on the gentler slopes and the lobes on steeper ones. The general form and steep fronts of the benches and lobes make them look much like the folds in an elephant's skin; see **Figure 4.31**.

57. Wahrhaftig (1949).

Figure 4.31B Top: The arrow in the photograph at upper right taken from the highway on Eagle Summit indicates the position of the bottom photograph and the photographs on the next page.

Figure 4.31C Top: Views to the west across the lobes. Bottom: View east toward the highway. In that photograph, the 1-meter scale dog digs for rodents resident in the toe of the lobe, and just below, arrows indicate the inferred solifluction motion.

The rates at which material soliflucts down a slope depend not only on the slope gradient but also on the availability of water, the distribution of grain sizes in the material, and perhaps the vegetation cover. Measurements on shallow to moderate slopes (5° to 20°) over a period of years indicate movements ranging from less than 1 cm/yr in dry areas to a maximum of 6 cm/yr in wet areas, with speeds near 1 to 2 cm/yr most typical.[58] Silty soils are the most prone to solifluction flow, more so than clays and far more than sands and gravels. (Recall that silty soils also favor the formation of segregation ice, the melting of which is thought to be important in promoting solifluction.) The role of vegetation cover is less clear in that it may impede solifluction by anchoring soil and drying it through plant respiration, or, especially if it makes soil humus-rich, the vegetation might contribute to solifluction flow by helping to retain water in the soil.

Rock glaciers

An active *rock glacier* is a distinctive flowing landform, typically several tens to 100 meters thick that forms on mountain slopes having a mean annual temperature low enough that water percolating into blocky soil debris can freeze into a long-lasting ice matrix. The material in the rock glacier may derive from that in a talus slope or from the debris deposited by a normal ice glacier at its terminal moraine, which is the case in the example of a rock glacier shown in **Figure 4.32**. Whatever its source, when this material develops a sufficient ice matrix within its internal voids, it takes up a life of its own, flowing downhill through deformation of the interstitial ice.[59] The slope-dependent rate of flow is typically in the range of 1 to 100 cm/yr, and a few observations indicate speeds exceeding 1 m/yr.[60] Thus, rock glaciers typically flow at rates near the upper range of solifluction flow (a few cm/yr), but at rates far less than ice glaciers in the same environment. Cold ice glaciers flow only a few meters per year, while those in relatively warm alpine settings flow at speeds exceeding 100 m/yr.

The top two or so meters of a rock glacier is essentially all rock fragments and perhaps some soil, but below this rind (essentially the active layer), ice makes up approximately one-half of the rock glacier's volume. It is down here that the flow occurs through deformation of this ice. The flow occurs mainly in summer and is faster the warmer the ice, so the

58. Washburn (1980) p 204.
59. Wahrhaftig and Cox (1959).
60. Vitek and Giardino (1987) p 16 (Table 1.1).

Figure 4.32 A large rock glacier fed by debris from a cirque glacier near McCarthy in Alaska's Wrangell Mountains. Photographed in 1979.

faster motion is in the upper part of the ice-rich body of a rock glacier. This motion carries along the topmost few meters. The material in that top layer generally tends to be more blocky than at depth; the rocks are angular and, depending on the nature of the source material, they may be mixed with a poorly sorted sandy or shaley matrix.[61]

The differential flow—faster near the top than at depth—creates the key distinguishing feature that allows easy identification of an active rock glacier: it has a steep frontal ramp breaking sharply away from the top surface of the rock glacier and inclined at the angle of repose of the material in the core of the rock glacier. The ramp of the rock glacier shown in **Figure 4.33** is a 32-degree slope, and the top surface of the glacier is roughly parallel to the ~20-degree mountain slope on which the rock glacier rests. Active rock glaciers maintain their steep frontal ramps because the faster downslope movement of the surface portion of the glacier delivers a steady supply of new material to the terminus where it cascades down over the slower-moving material beneath.

Figure 4.33A A small rock glacier photographed from the Richardson Highway at Mile 208 in 1996. The black dots mark the approximate upper boundary of the glacier, and the white arrow marks the location of the photographs in Part B.

---

61. Calkin et al. (1987) p 69; Barsh (1988) p 74.

Figure 4.33B At top left the 1-meter scale dog Foxy scrambles directly up the frontal ramp and, at right, slinks unhappily across it because she has discovered that every footstep can start a slide of the loose rock and soil lying at its angle of repose. Bottom, a view directly across the frontal ramp toward the less steep colluvial slopes beyond. See also Plates 14A and 14B.

The top surfaces of rock glaciers exhibit the same signs of flow seen in ice glaciers and in other bodies of viscously flowing material: lobate structures and transverse and longitudinal ridges and furrows, such as easily seen in Figure 4.32. Variations in topography, supply of material to its upper reaches and perhaps other changes can cause a part or all of a rock glacier to become inactive. Signs of that happening are the development of vegetation on the frontal ramp and substantial furrowing of the ramp by water erosion, as is evident on parts of the rock glaciers shown in Figures 4.32 and 4.33. Also, the surfaces of rock glaciers may show evidence of changing flow patterns that cause some parts of a rock glacier to override others that have slowed down or become inactive.

Rock glaciers are found in the near-glacial parts of all major mountain systems, from the Andes to the Alps (which alone contain more than 900) and from the Antarctic to the Arctic. Two hundred have been identified in relatively small parts of the Alaska Range.[62] Although rock glaciers do not transport material very far nor very fast, they are an important means of moving rock downslope, accounting for 15 to 20% of the periglacial transport in the Swiss Alps.[63]

Rubble sheets (block fields and block slopes) and rubble streams (block streams)

In situations where frost wedging of massively jointed bedrock produces blocky angular rocks, these may stand up out of an array of finer material that undergoes solifluction and transport by water or wind erosion.[64] Geocryologists generally give the name **block field** to **rubble sheets** on level or near-level ground, and **block slope** to those on slopes of 10° or more. They apply the terms **rubble stream** and **block stream** to linear arrays moving down the steepest slopes available or confined to narrow valleys. To qualify as a rubble sheet or stream, the deposit must have more than half of its surface (typically one to three meters in thickness) composed of blocky material ranging in size upward from 10 cm.[65]

Rubble sheets and streams differ from talus slopes in usually not having a cliff or steep angled source of rock at their heads. Instead, they derive their rubble through the wedging of blocks from bedrock by frost action, or in some instances they may lie downslope from till bodies that supply their blocks. The developmental history of many rubble deposits begins with cryogenic weathering and sorting of bedrock by frost action followed by solifluction flow downslope and removal of the finer-grained material

---

62. Wahrhaftig and Cox (1959).
63. Barsh (1988) p 85.
64. Most cryologists consider solifluction to be the primary transport mechanism but studies by Thorn (1988) indicate that erosion performed by flowing of snow meltwater can be appreciable.
65. Washburn (1980) p 219; Péwé and Reger (1983) p 125.

largely by water and wind erosion. On shallow slopes gelifluction and frost creep appear to be the primary transport processes, and on steeper slopes water runoff and rapid mass wasting (mudflows) become more important.

Some rubble sheets may be of fairly recent origin—one in Finland appears to be less than 6000 years old[66]—but ones in the Alaska Range are much older, forming during colder parts of the Wisconsin (75,000 to 10,000 years ago). In fact, in one Alaska Range locality, Jumbo Dome, a sequence of five rubble sheets has formed, each sheet probably representing a separate glacial substage of the Wisconsin.[67]

Ploughing blocks and braking blocks

Seen on slopes undergoing solifluction, *ploughing blocks* are large rocks that move 2 to 10 times faster than the surrounding soil because of enhanced gelifluction beneath the blocks.[68] They tend to turn so their long axis is aligned along the slope. Helping to identify them is the furrow each leaves behind, and perhaps a slight mound in front. A bit harder to recognize are *braking blocks*, large rocks that move slower than the surrounding solifluction, tending to impede the flow. Both ploughing blocks and braking blocks are widespread in alpine environments, more curiosities than anything else, but they do help mark slopes as undergoing or having undergone solifluction.

Cryoplanation surfaces

The shapes of eroded landforms seen in upland areas reflect the weathering and transport processes that created them: chemical weathering smoothes sharp edges and so tends to produce rounded hills; flowing water incises the hills to carve between them V-shaped valleys, and glaciers—like giant ice cream scoops—gouge out rounded cirques and U-shaped valleys. In cold rocky upland areas, the combination of cryogenic weathering and cryogenic transport can produce yet another characteristic shape, the gently inclined plane known as a *cryoplanation surface*. Cryoplanation surfaces are of two main kinds that share in common the basic planar shape but differ enough in size, location, and origin to deserve separate names: *cryoplanation terrace* and *cryopediment*. Table 4.1 summarizes the main characteristics of these two erosion features.

---

66. Washburn (1980) p 222.
67. Wahrhaftig (1949); Péwé and Reger (1983) p 60, pp 125–27.
68. Washburn (1980) pp 223–24.

Table 4.1 Characteristics of cryoplanation surfaces, a slightly simplified version of a table given by Priesnitz (1988) and incorporating some results of Reger and Péwé (1976)

Attribute	Cryoplanation Terrace			Cryopediment		
Location and relation to other topography	On hilltops or interrupting upper or middle slopes at several levels			At foot of mountain or valley slope, usually at one level only		
Morphology:	low	common	high	low	common	high
downslope width	5 m	100 m	>1 km	100 m	1 km	>10 km
alongslope length	30 m	400 m	> 10 km	1 km	10 km	>100 km
tread gradient	1°	6°	14°	1°	3°	12°
riser gradient	9°	30°	90°	7°	25°	35°
riser height	1 m	6 m	50 m	10 m	100 m	>300m
Original slope	25° or less			No restriction		
Original slope lithology	Generally in hard rocks producing coarse debris			Hard and soft mainly sedimentary rocks		
Tread material	Bedrock surfaces mantled by <1–3 m solifluction and outwash debris			Bedrock surfaces mantled by <1–3 m solifluction and outwash debris, fluvial gravel		
Riser material	Frost-shattered cliff, blocks and bedrock			Frost-shattered slope, talus mantle rock		
Structural control	Apparent in many cases			None, cuts across original structure		
Observed processes in upper part of tread	Frost wedging and shattering, solifluction, meltwater washing, piping			Frost wedging, sheetwash, solifluction, piping		
Observed processes in lower part of tread				Sheetwash, stream transport, cryoturbation, frost sorting		
Observed processes in the riser	Transverse wearing by nivation enhanced frost wedging and shattering, solifluction, meltwater washing, piping			Frost weathering, gravitational transport, rillwash. Mainly transverse wearing but also down wearing in soft rocks		
Origin of water at upper end of tread	Snowbank melting, seepage water			Runoff from upper slope, meltwater, seepage water		
Permafrost dependence	Yes (but controversial)			Yes		
Estimated formation time	10,000 years			100,000 years		

Except for some cryoplanation terraces that sit atop hills, the planar portions (treads) of cryoplanation terraces and cryopediments abut steeper slopes (risers) at their upper ends. These abrupt junctures are a distinguishing characteristic of all but the hilltop cryoplanation terraces. (They too had them once, but now the risers are cut away.) Structural irregularities in the parent slopes probably initiate cryoplanation terraces. Like a pothole in a road, an irregularity in a cold upland slope tends to grow, and the physical processes involved—frost action, solifluction, piping, and wind deflation, especially in a permafrost setting—cause the growth to be primarily transverse, that is, essentially horizontally into the slope. The growing notch makes a fine setting for catching windblown snow, and when the snow melts it provides a plentiful and continuing supply of water to augment the erosion and transport processes. *Nivation* is the name applied to relatively rapid erosion taking place at the edges and underneath a lingering snow bank, and the cavity it cuts is sometimes referred to as a *nivation hollow*. Thus, a nivation hollow can be thought of as the cutting edge of a cryoplanation terrace. The supply of moisture is critical to both the effectiveness of frost wedging and solifluction, and both surface and subsurface runoff help transport debris across the tread of the terrace. Because it forms a floor for the erosion and transport processes, permafrost is crucial to the growth of both cryoplanation terraces and cryopediments.[69]

Whereas cryoplanation terraces sit on hilltops or in bench-like arrays on the upper and middle portions of slopes, the much larger cryopediments lie in the bottoms of valleys or at the feet of mountain slopes. The supply of water crucial to continued widening of a cryopediment comes more from seepage and runoff from the upper slope of the riser than from the melting snowbanks that feed terrace erosion. Cryopediments tend to have slightly less gradient than cryoplanation terraces, and since they are observed to cut across topographic features their location depends little on structural control. Their risers are high, they undergo cryogenic weathering, and they readily shed the weathering debris downslope to the cryopediment where solifluction and water transport carry it large distances across the tread. The upper parts of the cryopediment treads look like treads of cryoplanation terraces, but the lower portions generally show more evidence of water transport across the surface, repeating ground patterns and more vegetation cover.

---

69. Reger and Péwé (1976); Péwé and Reger (1983) p 125.

The observed angularity of cryoplanation surface junctures with their risers testifies to high effectiveness of cryoplanation erosion processes relative to those (chemical and fluvial) that tend to round and incise topographic features. A cold, dry climate and a shallow permafrost table favor cryoplanation, whereas climate amelioration favors the rounding and incising processes. It follows that cryoplanation has been more active in the past, and in fact many if not most cryoplanation surfaces are now inactive.

Cryoplanation is a slow process, so most cryoplanation surfaces seen now (**Figures 4.34** and **4.35**) are the product of cryogenic erosion active for millennia, near 10,000 years for terraces and 100,000 years for cryopediments. By relating cryoplanation surfaces to sediments and river terraces of known age, or to known glaciations, it is possible to determine when the surfaces were formed and active. Some cryoplanation terraces studied in the Alaska Range are known to be less than 75,000 years of age,[70] and in the Richardson Mountains of Yukon Territory dating of moss on terrace treads indicates that active cryoplanation probably continued up to a few hundred years ago.[71]

Tors

As they near the end of their work, the processes that create cryoplanation surfaces sometimes leave monuments of their achievements in the form of *tors*, spires standing up on all sides above the surrounding slopes. Tors are the products of differential weathering, and they owe their existence mainly to differences in bedrock joint spacing. The weathering and debris removal processes are not necessarily cryogenic, and although tors are more common in polar and subpolar climates, they also are found in warm moist and tropical climates where deep chemical weathering and fluvial transport of debris are the key processes. However, frost wedging and cryogenic mass transport probably dominate over other erosion processes in producing the cold-climate tors. Shown in **Figure 4.36** is a group of tors accessible to hikers in the Fairbanks area of Alaska. These stand starkly above sloping terrain that probably is not a true cryoplanation terrace.[72]

70. Péwé and Reger (1983) p 126.
71. Priesnitz (1988).
72. Richard D. Reger, personal communication, January 1998.

Figure 4.34 A cryoplanation terrace on the north side of the top of Eagle Summit, Alaska. Notice the angular uplifted rocks on the terrace and the sharp intersection with the steeper slope above.

Figure 4.35 A cryopediment about 4 km wide and with inclination 2° to 4° in the Little Keel River Valley in northern Canada's Mackenzie Mountains. Notice the sharp juncture with the mountain slopes. Reproduced from Figure 3.3 of Priesnitz (1988).

Figure 4.36 Granite tors standing above a gentle slope in the foreground that shows evidence of sorted polygons. Photographed northeast of Fairbanks, Alaska, by A. F. Weber.

▪ Patterned Ground

Among the interesting things to look for and ponder over when traveling in any area are striking landforms and repeating patterns that reflect ongoing or past physical processes. A person cannot help but marvel over the power of the forces that have acted to create the repetitively folded rock strata seen on a rugged mountain slope nor, when viewing multiple, regularly spaced beds of coal sandwiched in between thick layers of sandstone, avoid asking just what curious set of circumstances could have caused this succession.

The contorted rock strata and the alternating coal and sandstone beds are the products of large-scale geologic processes initiated millions of years ago and which may still be acting. Also producing repeating patterns in the landscape are cryogenic processes—those involved in freezing and thawing—and they too may have operated only in the past or perhaps be still ongoing. However, the time scale is shorter—days to hundreds of thousands of years instead of millions of years—and the spatial scale is also much less, ranging from a few centimeters to upwards of a few meters.

The term patterned ground generally refers to a land surface that exhibits repeating geometric shapes generated by processes acting within a few meters of the surface. In the context here, "patterned ground" refers to a land surface with patterns created by or at least influenced by cryogenic processes. Some of these patterns may be identical to or similar to others observed in warm climates but which are not cryogenically caused.

Observed patterns on the ground surface are there either because they show up as topographic features or as spatial differences in composition of the ground surface, and perhaps in vegetation cover. Thus the patterns either are the products of processes that move soil in nonuniform ways or sort it according to size and thereby juxtapose small- and large-size particles in certain geometric arrays. In general, closed patterns—circular or polygonal shapes—develop on level ground, whereas open patterns—parallel or quasi-parallel banding or topographic irregularities—develop on slopes. Some of the closed and open patterns exhibit sorting, and some do not.

Examples of closed, nonsorted patterned ground created by cryogenic processes are the widespread arrays of high- and low-center polygons previously discussed and which are generated within perennially frozen ground. These polygonal forms are large, ten to several tens of meters across, and each has taken hundreds of years to develop. Many of the ice wedges forming the skeletons of these arrays are relics of past times when the climate was colder; many are no longer growing but remain perfectly preserved within the permafrost, and yet others—those associated with high-center polygons—are now melting away because of lowering of the permafrost table for one reason or another.

Most other cryogenically generated ground patterns are due to processes within the active layer, but they may be influenced by what lies below. Patterned ground developed in a setting where the active layer is in contact with the permafrost table may differ much from that generated where no permafrost lies below, or where a layer of thawed soil (talik) separates the bottom of the active layer and the permafrost table. If in contact with the active layer, the permafrost table in cold areas such as northern Alaska and Canada can have a temperature well below 0°C during part of each summer. In this situation, upfreezing can proceed near the bottom of the active layer simultaneous with thawing of its upper part, and in fine-grained soils the associated suction can draw water down from above and generate frost heaving near the base of the active layer.[73] During the upfreezing, any horizontal irregularities that affect thermal conductivity or water flow are likely to produce differential heaving and discontinuous deposits of segregation ice deep within the active layer, in the same fashion as downfreezing produces them in the upper part of the active layer. Owing to its relative impermeability, the permafrost table also acts as barrier to downward flow of water, and it, compared to a wet talik layer, is a poor source of moisture which downward freezing might try to draw up into the active layer.

Local climate also is an important determinant of patterned ground. A temperate climate that allows frequent cycling of temperature through the freezing point fosters near-surface cryogenic processes such as pipkrake growth, a process believed important to the generation of some of the open patterns that develop on slopes. In a colder climate the more extreme annual temperature variation fosters deeper cryogenic processes such as frost heaving, ground cracking, frost wedging, and thaw consolidation, all of which can influence ground patterns. Vegetation plays a role as well by stabilizing soils on slopes, altering the thermal conductivity of the top layer, affecting the supply of water at the surface, and interacting with surface irregularities in ways that affect both the plant distributions and the irregularities. Thus the distribution of related plants can become an important part of whatever ground pattern develops, perhaps even its most visible component.

---

73. Mackay (1980).

Insight into the processes creating patterned ground comes largely from examination of the patterns in the field, field experiments, and laboratory experiments. It is a complex matter, so interpretations differ and in some cases the conclusions are more aptly described as conjectures. One widely accepted idea is that thermal or desiccation cracking initiates some kinds of patterned ground, and determines the spacings and mesh sizes of the elements in a soil pattern. Thermal contraction cracking is the accepted explanation for the origination and configuration of ice wedge polygon arrays. More widespread, especially in fine-grained soils, desiccation contraction cracking occurs when water leaves a soil through evaporation, gravity drainage or the cryogenic suction that develops during freezing. Both the thermal environment and the strength of soil involved are likely to be important in determining the space and depth of contraction cracking, regardless of cause. The cracks then constitute structural and thermal irregularities in the soil structure that affect other processes involved in creating patterned ground. Apparently playing a major role in the generation of patterned ground is differential frost heave, the uneven displacements of frozen ground in consequence of differing grain size, temperature distribution and moisture content. Differential frost heave appears to be the main cause of cryoturbation (see earlier discussion), and it can separate out soils of differing heaving abilities horizontally as well as vertically, thereby causing them to lie side by side.[74] Contributing also to small-scale sorting and circulatory motions may be cryostatic pressure created within the thawed parts of the active layer sandwiched in between the frozen parts above and below. Thaw consolidation due to loss of water drawn away by the suction accompanying freezing of adjacent soil layers also can contribute to the uneven motions that develop patterned ground.

Nonsorted circles (also called mud circles, mudboils, earth circles, medallions)

Nonsorted circles are bare soil patches 0.5 to 3 m across that contain mainly silt and clay (see **Figure 4.37**). They are widespread in alpine, subpolar, and polar environments, in both permafrost and nonpermafrost settings.[75] The surface of a nonsorted circle is either flat or slightly domed, often dry and surrounded by lichens, but the fine-grained material at depth may contain enough water to be nearly at the ***liquid limit***, the point

---

74. Williams and Smith (1989) p 162.
75. Williams and Smith (1989) p 163.

Figure 4.37 Nonsorted circles (mud circles) near the Denali Highway near Mile 40 (top), and near Mile 11 (bottom). Photographed 1996.

at which the soil contains so much water it begins to behave as a liquid. In late summer rainwater or water released from melting in the active layer may increase internal pore pressures enough to cause liquefaction and a boiling up of the fines through the surface simply because of the stress from weight of overlying material. The names *mud circle* or *mud boil* then become quite properly descriptive. During spring breakup, vehicles rolling along roadbeds containing silts and clays similarly will cause mud to boil up through breaks in the road surface.

At depth, the fine-grained material forming the body of the material in a nonsorted circle may extend beyond the area of the circle itself, and excavations have in at least one instance shown interconnections with other nearby circles. In this case gravelly sand surrounded the silty clay body below the circles,[76] so here there seems to be evidence of horizontal sorting by differential frost heave and related processes.

Examples of a somewhat different pattern lacking sorting, the *nonsorted polygon* appear in **Figure 4.38**. This pattern, currently developing in a water-saturated, uniformly fine-grained river delta deposit, illustrates the major role that thermal or desiccation cracking can play in the development of patterned ground, and perhaps in the formation of hummocks as well, since these are forming in the driest parts of this deposit; see Figure 4.38E.

Sorted (stone) circles, sorted (stone) polygons, debris islands, and stripes

Cryogenic sorting of soils creates *sorted circles* and *sorted polygons* consisting of a central area of fines grading out to coarser rocky material; these are characteristic of level ground in permafrost areas. They are also present in subpolar highlands with a thick active layer. They range in size from as little as 10 cm up to about 5 m, the larger circles and polygons tending toward larger stones contained in their outer parts. If tabular, those stones tend to sit on edge and be aligned with their long axis parallel to the boundary of the circle or polygon.

The sorted arrays shown in **Figures 4.39** and **4.40** contain examples of both sorted circles and sorted polygons juxtaposed in such a way that it is fairly obvious that the arrays began as polygons, and then as their development ensued the trend is toward becoming more circular. Thus age

---

76. Washburn (1973) pp 103–06.

Figure 4.38 Thermal and desiccation cracking creating polygonal patterns in saturated, fine-grained soil deposited by Slims River where it empties into Kluane Lake at Mile 1028 on the Alaska Highway. Parts A and B show large parallel cracks spaced approximately 2 m apart with an inner array of polygonal cracks of spacing approximately 0.7 m, as shown in Parts C and D. The photo in Part E shows what appears to be a group of earth hummocks beginning to form. These are approximately 20 cm in height. Photographs taken August 1996.

Figure 4.39 Sorted polygons at Mile 31.8 on the Denali Highway, Alaska. Top: Lightly vegetated soil showing thermal or desiccation cracks at lower-right corner grades into the sorted polygons at center, and these grade into the rock pavement beyond, the sequence illustrating the transition with time. The close-up views at bottom show newly exposed white rocks in the central part of polygons. (The two shown in the right-hand picture are the two nearest the camera in the left-hand picture.) Notice that the rocks grow darker the longer they are exposed, and that the boundaries between the inner fines and the outer rocks are trending toward circular. Photographed August 1996. See also Plate 15.

Figure 4.40 Sorted polygons at Mile 37 on the Denali Highway. This array appears in the foreground of Figure 4.21 Here the individual polygons are up to 5 m in diameter. Photographed in August 1996.

more than mechanism would appear to determine if a sorted array is polygonal or circular.

Excavations into sorted polygons and sorted circles indicate that the size of the stones in the outermost parts often but not always decreases downward.[77] Sorted circles and sorted polygons develop on slopes up to 30° where they tend to be more irregular and hence have received the name *debris islands*.[78] **Figure 4.41** shows a spectacular array of sorted circles, and it is obvious that these are quite different from the sorted polygons depicted in the previous two figures, and perhaps the mechanisms involved differ.

---

77. Washburn (1973) p 123.
78. Washburn (1980) pp 129–30.

Figure 4.41 Dramatic examples of sorted stone circles photographed in Spitzbergen by Bernard Hallet (upper photo) and Alfred Jahn (lower photo).

Observations of sorted circles and sorted polygons formed in glaciated areas and exposed in recent decades or centuries coupled with results of experiments involving removal of surface vegetation indicate that sorting by differential frost heave plays the major role in forming these geometric patterns, and that desiccation and thermal cracking contribute significantly.[79] Observations made for several years after intentional stripping of vegetation from the surface of an unsorted soil, show that the larger rocks move rapidly toward the surface.[80] If any desiccation or thermally induced cracks appear at the ground surface, they create additional transverse temperature gradients that cause rocks to migrate toward them. The frost heaving also orients the rocks as they move outward, leaving the smaller particles near the center. In addition to thermal or desiccation cracking, processes triggering the sequence ultimately leading to a sorted polygon and then later a sorted circle might be initiated by any irregularity in the ground surface that affects the supply of water or fosters creation of horizontal temperature gradients. The falling over of frost-wedged stones, alteration of vegetation by browsing caribou and small rodents or erosion by water or wind are among the possibilities that may augment desiccation or thermally induced cracking. Once the sorted polygon-building sequence starts, it seems to perpetuate itself as long as conditions are favorable or until the growth itself modifies the situation enough to terminate further growth. One geocryologist has suggested that only those regions with mean annual air temperatures less than –4°C and having ice-rich soils produce well-developed sorted circles and sorted polygons of size greater than 2 m.[81]

Circular or polygonal patterns obviously resulting from cryogenic processes include small (decimeter-scale) configurations involving arrays of small stones surrounded by fines—just the opposite of stone circles—and networks of small arrays within larger ones. Considering the variability in soils and climatic conditions, it is not surprising that the observed variety in cryogenically caused microrelief seems endless.

---

79. Desiccation or thermal cracking obviously is the origination of the stone polygon nets shown in Figures 4.38 and 4.39 since it is easy to see the progression from the beginnings of thermal or desiccation cracking on one boundary of each array to the final product on the other, a block pavement containing no fines at all. These net arrays formed in a small drying lake bed in a glacial moraine. Observations of them and other net arrays in September 1996 and August 1997 showed much difference in water levels: some were completely exposed in 1996 but under several centimeters of water in 1997. In nearby shallow lakes some nets could be seen a meter or more below the water surface.
80. Pissart (1974) cited by Williams and Smith (1989) p 167.
81. Goldthwait (1976); Washburn (1980) p 145.

Closed sorted-microrelief patterns are restricted to level or near-level ground. On even rather slight slopes, those of only 3° or 4°, the closed patterns may become elongated parallel to the slope and develop into *sorted stripes* ranging in width from as little as 10 cm to more than 1 m. Sorted stripes consisting of alternating bands of fines and coarse materials propagate themselves downslope at least in part through differing rates of movement dependent on the size of the particles involved. Observations indicate slow movements ranging from a fraction of a cm/yr to more than 10 cm/yr; sometimes the coarse bands move faster than the fines, but usually the bands containing the fines move the fastest.[82] The long axis of stones in the sorted strips tends to be in the vertical plane and oriented parallel to the slope, and the size of stones usually decreases rapidly with depth. The faster transport of fines by water in summer may be part of the cause of differential speed, the tendency of larger stones to inhibit solifluction another, and the ratcheting by pipkrakes (needle ice) of the top layer above buried stones yet another part of the cause. The tendency for pipkrakes to occur in soil already loosened by former pipkrakes and to draw water and perhaps fines horizontally as the ice needles form may also contribute to the sorting that originates the stripes.

Nonsorted stripes also exist, consisting of alternating vegetated and nonvegetated bands or merely alternating bands of contrasting-colored exposed surface material. The stripes shown in **Figure 4.42.** appear to be of this type. Again, differences in rates of downslope movement caused by density differences, water retention, and vegetative binding of soil probably perpetuate these stripes as well.

Cryoturbation steps (terracettes) and sorted steps (stone garlands)

Cryoturbation steps, also called terracettes, are remarkably regular arrays of small benches 30 to 80 cm wide with risers between them of approximately the same size. **Figures 4.43A-E** show examples; see also Figures 3.3A and B. They can be seen on steep, lightly vegetated slopes in many parts of the northern half of the United States and in Canada. The steps are poorly sorted or nonsorted features that develop only on lightly vegetated hillsides with slopes in the range of 20° to 38°, the risers tending to be vegetated and the treads bare. Pipkrake formation is thought to play the primary role in generating the cryoturbation steps by lifting the smaller soil material up away from buried pebbles or rocks, or by lifting fully exposed rocks and shunting them

---

82. Washburn (1980) pp 151–54.

Figure 4.42 Stripes on hills composed of glacial till near Mile 40 of the Denali Highway. Brushy vegetation anchors the soil between the lighter colored strips. Photographed August 1996.

downhill, thereby creating some sorting.[83] The snow lying on the treads of the steps shown in Figure 4.43D suggests also that nivation (enhanced erosion at the edge of a snow deposit) on a microscale might play a part in enlarging the steps, and perhaps even provide an explanation of why the

83. Tricart (1970) pp 94–95; Washburn (1980) pp 147–51.

Figure 4.43A
Cryoturbation steps approximately 10 km north of Cache Creek, B.C., on Highway 97. Top: view along the cryoturbation step identified by the arrow in the middle photograph (taken simultaneously by Rosemarie Davis), and at bottom, a more distant view of the array, looking south. Photographed in 1996. See also Plate 16.

Figure 4.43B Large cryoturbation steps approximately 1 m wide on south-facing hillsides 13.6 km north of Cache Creek, B.C., on Highway 97. Photographed in 1998.

steps typically are horizontal or nearly so: once nivation starts in a snow-filled depression, it could enlarge the depression horizontally along the hillside as well as into the slope. Also, in a wind-blown area the steps would tend to catch snow, and the melting of the snow should add water to the soil on the step treads, perhaps fostering the growth of pipkrake ice during periods of freezing. Wind may have a role here too, helping to remove fines brought to the surfaces of step treads by pipkrake action. All this is conjectural, and it does not seem to give a reasonable explanation for the sizes of the steps and why they are so regular. Seemingly possible causes of the spacing between steps are local depth of freezing, soil depth, and layering within the soil or other characteristics that might affect how far cryosuction on the treads can move water horizontally from beneath the risers. Steps occur only on generally north-facing slopes in the more southerly areas where they appear, evidence that microclimate surely enters much into it too. Like the ripples on sand dunes, cryoturbation steps seem to be another example of self-organizing forms that start out as small bumps or hollows and evolve into larger forms having regular patterns as they migrate downslope.[84]

---

84. Werner (1999).

Figure 4.43C Approximately 90 cryoturbation steps on an east-facing hillside approximately 10 km north of Hudson Hope, B.C., on Highway 29. Photographed in 1998.

Figure 4.43D Horizontal cryoturbation steps photographed in March 1999 when snow was melting, with an off-road vehicle track cutting across them from center toward upper left. These steps are located 2 km east of those shown in Figure 3.3A. The approximately 100 in the array extend from the base to the top of the north-facing hillside.

ANIMAL TRAILS ⟶ POORLY DEVELOPED PARALLEL STEPS
Interstate 82, Prosser, WA, 3 Miles West ⟶

Figure 4.43E Animal trails radiate from a watering tank up across an array of weakly developed cryoturbation steps that are more fully developed on the hillside to the right of the photograph taken looking south in March 1999.

At first glance a person might attribute the cryoturbation steps to the action of grazing animals, but that is not a viable explanation. It is true that once the cryoturbation steps have formed, grazing animals sometimes do tend to follow along them, so the animals may contribute secondarily to the growth of certain of the steps. That is most noticeable near fences where the frequent passage of cattle or other animals along the steps tends to widen the treads and leave enough fertilizer to create lush vegetation on the risers, making those selected steps stand out from others nearby. Animal trails sometimes cut across the steps, as shown in Figure 4.43E. This photograph helps show the distinction between cryoturbation steps and animal trails: the trails tend to diverge irregularly from a feed lot or watering place, whereas the steps parallel one another in very regular fashion.

Another step-like configuration, *sorted steps* (also called stone garlands), appears only on shallower slopes (5° to 15°) and has fines in the tread areas and stones on the risers. Sorted steps typically look less regular than cryoturbation steps and may be elongated downhill with steps wider than risers. They may be intermediate forms between closed patterned

ground configurations and stripes, and thus be more related to solifluction than to pipkrake formation.

■ Thermokarst Landforms

The term thermokarst refers to topographic depressions created by thawing of ground ice. Climatic warming is one reason thermokarst topography develops, and another is localized disturbance to the ground's
thermal regime that promotes thawing of permafrost. Forest fires, shifting
stream channels, and the activities of man are primary causes of rapidly
developing thermokarst topography. Then too, as part of their normal life
cycle some cryogenic landforms—such as palsas and pingos—develop
thermokarst collapse features. High-center polygon arrays constitute
another form of thermokarst topography, be they the result of climatic
warming or localized disturbance such as occurs from the clearing of
ground cover for construction and agricultural purposes.

Thermokarst lakes and beaded streams

Lakes and streams are always in transition. Streams continuously erode
some parts of their beds and deposit material in other parts, and most
lakes, especially those in formerly glaciated areas, grow smaller as they fill
with sediments and encroaching vegetation. Exceptions to that general
trend are **thermokarst lakes** (also called thaw lakes) which grow because
of ice melting beneath and around them. That melting occurs because
some change has taken place to alter a preexisting situation wherein permafrost was on the increase or stable. At nearby locations similar changes
might or might not have occurred. Examples of changes likely to cause
highly localized thawing of ice-rich permafrost are the normally continuous shifting of meandering streams in their beds, increasing inflows of
water to lakes because of changes in streams that feed them, and slumping
or settling caused by earthquakes or slower tectonic changes. In areas
where permafrost thawing is widespread, the most likely change is a
warming in regional or global climate.

One characteristic indicator of change in areas containing polygonal
arrays of ice wedges is the development of **beaded stream** patterns consisting of sharply defined pools of water interconnected with short drainage
channels that usually are straight or composed of straight segments con-
nected by sharp angles. Several examples appear in **Figure 4.44**. The

pools, typically 1 to 3 m deep and up to 30 m across, form mainly at the intersections of melting ice wedges, and the connecting channels tend to follow ice wedges. Beaded streams are distributed over widespread parts of northern Alaska, and are often found in conjunction with pingos and thaw lakes.

Steep collapsing banks creating treacherous shores and, in forested areas, inward-leaning trees signal thermokarst lakes. They are so common in the discontinuous permafrost zone of Alaska and Canada that most ponds and small waterbodies there are likely to show these thermokarst lake features. Examples appear in Figures 1.4 and **4.45.**

Oriented lakes

On Alaska's arctic coastal plain, near Point Barrow, groups of oriented thermokarst lakes create a spectacular scene when viewed from the air. Other groups of oriented lakes—those that tend to be elliptical or subrectangular and having parallel alignments—occur elsewhere, to the east of the Mackenzie Delta, on the Old Crow plain of northern Yukon Territory, and on Baffin Island, but the Point Barrow group is the most striking.

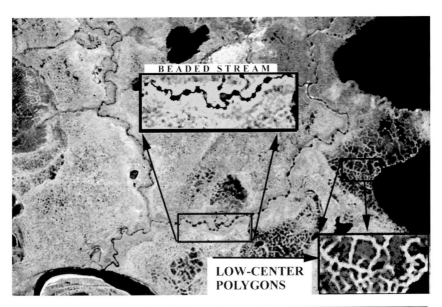

Figure 4.44 Beaded stream drainage and low-center polygons in a high-altitude infrared photograph of Alaska's North Slope. See also Plate 17.

Figure 4.45 Two views of thaw lakes bordered by tilting and submerging trees, signaling that the lakes are undergoing rapid enlargement. The upper photograph was taken in September 1996 at Mile 1125.5 on the Alaska Highway and the lower along the highway some miles to the north, also in September 1996. See also Plate 18.

The oriented lakes near Point Barrow are elongated in the direction 10°–15° west of true north, essentially at right angles to the prevailing wind. Intuitively, a person would expect that if the wind had anything to do with the elongation and orientation of the lakes, that elongation would be parallel to the wind. Thus the first speculations on the cause of the oriented lakes were either that the wind was not the cause or that the prevailing winds must have shifted by 90° some time ago. Careful investigations carried out on the Point Barrow oriented lakes seem to have dispelled these earlier thoughts and also counter more recent suggestions that the orientation might be determined by geologic structures rather than wind-related processes.[85]

The oriented lakes are shallow, never exceeding 3 m. Those less than 2 m deep freeze to the bottom each year, but deeper ones do not, and those modify the depth to the top of the permafrost layer below. Beneath one lake 3 m deep, the depth of the permafrost table was found to be 60 m, whereas elsewhere in the area it is 0.3 to 0.6 m (the bottom of the permafrost in this area is at depth approximately 400 m).[86] Since many oriented thermokarst (thaw) lakes occur in association with polygonal ground patterns, they are thought likely to originate much like beaded drainage ponds, but in locations without comparable drainage. Many of the oriented lakes lie partly or wholly within the confines of former lakes reclaimed by sediments and vegetation.

Once initiated, the lakes grow preferentially at right angles to the wind because wind-driven wave and current patterns cause the fastest thawing and mechanical erosion at the cross-wind ends of the lakes. In the Point Barrow area the prevailing summer wind is from the east-northeast; at other times a second lesser prevalence is from the opposite direction, south-southwest. In the early stage of a lake's development, the wind-driven wave action tears sunken mossy vegetation from the lake floor and carries it toward the west and east shores. That action removes insulation from the floor of the lake, allowing deeper thawing and subsidence there, and deposits it in near-shore shelves and bars where it inhibits wave erosion and thawing. This transport of bottom material parallel to the wind is only part of the story. The rest has to do with subsurface current patterns set up by the prevailing winds. The key feature of these patterns, shown in **Figure 4.46**, is a fast flow of surface and subsurface water at the crosswind ends of the lake, at speeds typically several

85. Carson and Hussey (1962); Britton (1967); Sellman et al. (1975).
86. Brewer (1958).

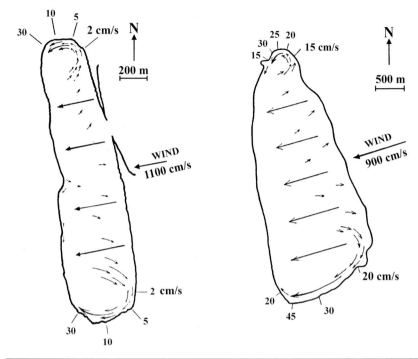

Figure 4.46 Wind and current patterns observed in two aligned lakes near Point Barrow. The diagram is compiled from parts of Figures 3, 9 and 10 in Carson and Hussey (1962).

times the subsurface flow elsewhere. With a 32-km/hr (20-mi/hr) wind, the currents at the north and south ends of the larger lakes reach speeds near 1km/hr (2/3 mi/hr) and that is fast enough to move even pea-size gravel along the beaches at speeds approaching 100 m/hr. This ability to transport material away from the ends, coupled with the ability of the faster moving water to deliver heat energy for thawing, evidently combines to foster erosion of the north and south shores.

High-altitude photographs of the oriented thermokarst lakes near Point Barrow reveal a striking pattern of sublittoral (nearshore) plant growth in those lakes now growing smaller, and the photographs show that this growth can accentuate the elongation of each lake as its area contracts. Tentatively identified as primarily *Arctophylla fulva*—a grass that grows in water up to depths of 1.5 m in arctic regions[87]—the vegetation grows outward

---

87. Viereck, Les A., personal communication, 1997. Britton (1967) gives information on this and related grasses that grow in shallow arctic waters.

from the shore in pronounced circular patterns that eventually coalesce to create a new shoreline subparallel to the former one. The photographs in **Figures 4.47** and **4.48** indicate that the new shores extend as far as or even farther out on the west and east shores of the lakes than on the north and south shores, and so the lakes become more linear as time progresses.

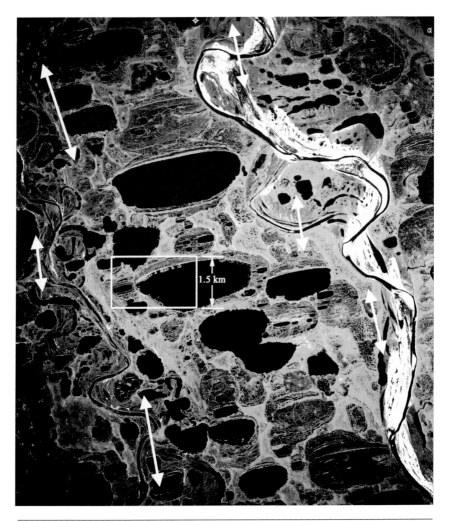

Figure 4.47 High-altitude infrared photograph of aligned lakes on Alaska's North Slope near Point Barrow. White arrows indicate approximate alignments of nearby snowdrifts, seen in the photograph. The portion within the white box is enlarged as Figure 4.48. See also Plate 19.

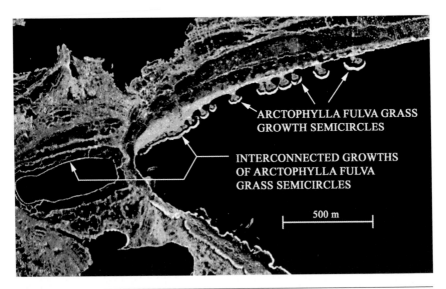

Figure 4.48 An enlarged view of the area shown in the white box of Figure 4.47. Around the small lake at left former shorelines are evident, each apparently created by joined semicircular growths of vegetation (Arctophylla [also Arctophila] fulva grass). See also Plate 20.

▪ Earthquake Effects in Permafrost Areas

During earthquakes, the intensity of motion of the ground surface generally is much greater in areas of unconsolidated sediments than in areas of rock. Because freezing bonds unconsolidated sediments, it is likely that those frozen sediments in areas of thick continuous permafrost will behave like rock and therefore undergo less motion than unfrozen sediments when earthquakes occur. In an area of discontinuous permafrost, however, the major differences in behavior of frozen and unfrozen bodies of soil can create ground motions of unusual sorts, generating spectacular earthquake effects not seen elsewhere. **Figures 4.49** to **4.53** present examples generated by a series of earthquakes in northwest Alaska on April 7, 1958; the largest event of this series was Richter magnitude 7.[88] Occurring when the active layer was frozen, these earthquakes produced visible secondary ground fractures, massive flows of sand and water, and significant ground surface collapses in a 15- by 60-km area of the Koyukuk River valley near Huslia. The sloshing

88. Davis (1960).

Figure 4.49 Mud thrown up by an earthquake darkens lake ice fractured by severe ground motion and sloshing of water below. Many lakes in the earthquake area showed such effects.

of water in the many small lakes in the area caused fracturing of the surface ice, piling it up on the shores and pumping portions of the lake bottoms up through the ice fractures, as shown in Figure 4.49. Similar effects occurred on old lake beds having thick moss covers or in swampy areas, where the ground was frozen on top but unfrozen below. The frozen surface broke into rigid blocks that slid across the unfrozen soil beneath and then piled up near the lake or swamp boundaries (Figure 4.50). Occasionally, unfrozen material below squirted up through the fractures between the blocks, as shown in Figure 4.51. However, the most spectacular consequence of severe

Figure 4.50 Water sloshing under the moss cover of old lakes during the earthquakes has cast these moss blocks up on the edges of the former lake beds. This and following photographs taken in April and early May 1958.

and prolonged ground shaking in this region of discontinuous permafrost was the ejection of many massive flows of water and sand onto the ground surface, much of this material having been transported through unfrozen subsurface channels prior to ejection. Accompanying the ejection of sand and water was the formation of surface collapse pits. One of the largest is shown in Figure 4.52; it was associated with a flow of sand with estimated

Figure 4.51   Rosemarie Davis, the author's wife and field assistant on a 160-kilometer investigative trek through the roadless epicentral area, trusty .30–30 strapped to her back to fend off any bears that might appear (none did), stands with one foot on a slab of mud forced up through a crack in the mossy active layer covering a swampy area affected by the earthquake.

volume 100,000 cubic meters. Figure 4.53 shows both the surface of one of the mud flows and trees pushed over by the sliding of portions of the frozen surface layer over unfrozen material below. These secondary earthquake effects stemming from the presence of partially frozen soils all were located in a totally uninhabited area; had they not been, the monetary damage could have been substantial.

Figure 4.52 A collapse pit 32 m across and 4 m deep associated with a mud flow having estimated volume 100,000 cubic meters. A smaller collapse pit is at the right-hand edge of the photograph. Notice the linear collapse feature marked by a dashed line that suggests underground flow of material from the small pit to the edge of the large pit where the huge amounts of water and soil flowed out onto the ground surface.

▪ Chapter Summary

In those parts of the world where the ground freezes annually and perhaps contains perennially frozen ground, cryogenic action causes weathering and generates certain geomorphic features. Some of these are similar to features produced in warmer climes by other processes, and some are unique to cold lands. The largest and most spectacular cryogenically caused landforms are open- and closed-system pingos, mounds up to 50 m tall created by the transport and freezing of water under hydrostatic pressure. Tending to be somewhat smaller are palsas that form where strong winter winds help generate strong thermal gradients conducive to the formation of thick layers of segregation ice which lift the ground surface. Cryogenic action also creates a variety of lesser mound forms ranging down to a few centimeters in size, and an almost bewildering variety of visible ground patterns. Of these, the most widespread and important in cold lands are polygonal ice-wedge arrays formed over long periods and which,

Figure 4.53 Rosemarie Davis, now carrying her ever-present bear gun on her shoulder, sashays over mud that has flowed up onto the ground surface near where shifting blocks of the frozen active layer have piled up and pushed over black spruce trees.

when melted, generate thermokarst topography. Other patterns involve sorting and short-distance transport of soil through cryogenic action. In some cold areas, cryogenic action's ability to break rock and soil down into smaller pieces and to move soil down even very gentle slopes (the solifluction process) helps make it the most important weathering and transport process. Repeated freezing and thawing of soil tends to sort particles by size, and to modify the shape, size, and packing of soil particles. Because of spatial inhomogeneities in the soil and in the thermal regime, repeated freezing and thawing tends to churn the soil as well, destroying the uniform layering usually seen in a soil profile generated by noncryogenic processes. Because frozen soil is brittle relative to unfrozen soil, strong ground motions can produce unusual effects directly attributable to differences between the motions of frozen and unfrozen ground. The earthquake effects include abnormal fracturing of the frozen active layer, sliding of frozen portions of the ground over unfrozen portions, and ground collapses associated with the ejection onto the surface of soil and water owing to differential motions between frozen and unfrozen soils.

Coping with Seasonal Frost and Permafrost

■ Early Northern Inhabitants and Permafrost

The Eskimos and other long-time residents of northern lands learned long ago how to coexist with the freezing and thawing of the active layer and with the perennially frozen ground often directly in contact with it. Because their livelihood came mainly from the sea and nomadic hunting, and they neither tilled the land nor mined it, the ground surface served these people mainly as an operating and living platform that they disturbed little—and which in return gave them adequate stability.

Along Alaska's northwestern and northern coasts, permanent settlements developed at places such as Point Barrow that were favorable to whaling. Pollution was not much of a problem in these villages because each winter any refuse placed outside the homes froze solid and some of it stayed frozen in summer, helping to build midden piles that grew over the centuries and raised the coastal villages above their surroundings, making it easier to spot whales right at the doorway. The people built winter homes from whale bones, driftwood, and sod, and they dug these into the ground. In summer they often lived in skin tents, especially if they were occupied with fishing at nearby streams and lakes. **Figure 5.1** shows such a camp with its tent frames sited on the driest places around, the centers of raised-center polygons. For use as food storage caches, the Eskimos excavated into pingos and pods of segregation ice. They buried their dead in permafrost (or in graves that soon became permanently frozen), and the frozen bodies became mummified by the same desiccating process that

Figure 5.1 An Eskimo summer camp with facilities placed on the driest places, the middle parts of high-center polygons (see arrows). The trail is made by dog sleds run across the tundra in summer. Photographed near Point Hope, Alaska, in 1957. See also Plate 21.

causes freezer burn in poorly wrapped frozen meats. One unintended result is that, when examined, the bodies have yielded much information about Eskimo life during past times.

■ Mining and Permafrost

Commencing in the 1870s, prospectors entered northwestern Canada and Alaska where their activities brought them up short against the hard reality that most of the gold they sought was buried in permafrost. They could avoid the problem by panning and sluicing the unfrozen ground in stream bottoms, but few would make much money that way. Not only did the early miners have to thaw the gold-bearing gravels, in most cases they had to thaw down through several to several tens of meters of frozen silty soil that overlay them. The main technique was to fight ice with fire, a slow process entailing a few hours of burning wood in a pit, excavation of the thawed material, and then a repeat of the cycle day after day. Another thawing technique used in the Fairbanks mining district of Alaska involved dropping heated boulders down the shafts.[1] Some of the resulting shafts in that district were as deep as 50 m. In the process the miners cut so many trees for shaft structures and firewood that they literally destroyed the nearby forests. Photographs taken of the early-day workings near Dawson and in the Fairbanks district show completely denuded hillsides in the background. (The trees have grown back, as the modern-day photograph of one of these shafts in **Figure 5.2** illustrates.)

Figure 5.2 The remains of an early-day mining shaft west of Ester, Alaska. See also Plate 22.

1. Forbes, Robert D., personal communication, 1998.

For prospecting purposes, a highly fuel-efficient innovation was the prospecting boiler (**Figure 5.3**) manufactured in the machine shop of the Northern Commercial Company in Fairbanks and perhaps elsewhere. Light enough to be carried by two men, the boiler generated enough steam to drive a 2- to 3-cm-diameter steam point down through frozen ground at rates exceeding one meter per hour and to thaw the soil within a few centimeters of the point.

Fire and steam were totally inadequate and too expensive to thaw the vast amounts of soil overburden and gold-bearing gravels involved when large-scale dredging operations commenced during the early 1920s in the Nome and Fairbanks areas of Alaska. Cold water, that with temperatures ranging from 0°C to 12°C, became the main thawing agent, and the procedures were sufficiently complex and valuable that their main developers,

Figure 5.3 A prospecting boiler rigged to burn fuel oil instead of wood.

subsidiaries of the United States Smelting, Refining & Mining Company (USSR&M), kept some secret and patented others.[2]

One simple device long used for thawing permafrost and removal of silt and clay overburden is the hydraulic giant, in essence a water cannon. Shooting out water under a head of 10 Atm, each giant can melt and carry away the overburden in a circular area 150 m in diameter. Twenty-six giants similar to that shown in **Figure 5.4** were in operation in 1926 on

Figure 5.4 A hydraulic giant used for thawing permafrost and washing away unfrozen overburden from gold-bearing placer gravels.

2. Wimmler (1926).

Goldstream Creek at the site now occupied by the town of Fox, just north of Fairbanks. They thawed and removed several million cubic meters of ice-rich silty overburden in the course of one summer.

After removing overburden with the hydraulic giants, the next step was thawing the gold-bearing gravels so that dredges could dig them. Strings of pipe distributed cold water under low pressure (1–2 Atm) to arrays of small-diameter (2–3 cm) pipe either hammered into the ground by manpower or inserted in holes drilled by sled-mounted churn drills. See **Figures 5.5–5.7**. With spacings ranging from 2 to 10 m, the USSR&M subsidiary had between 12,000 and 15,000 of these thawing points oper-

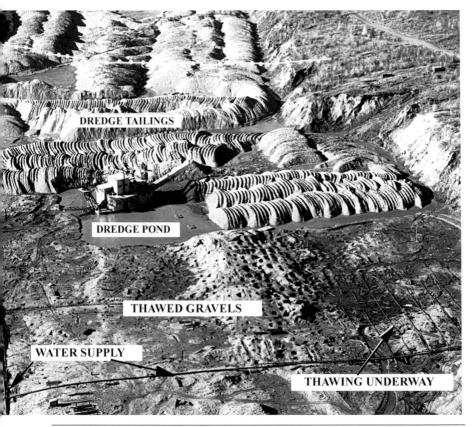

Figure 5.5 An operating gold dredge near Ester, Alaska, photographed during the early 1950s. The white dots indicate the location of the surface prior to mining. University of Alaska Fairbanks Archives, Bradford Washburn Collection, Photo No. 2407.

Figure 5.6 A closer view of the cold-water thawing activities shown underway in Figure 5.5. Circular dark areas are depressions created by melting of ice around each thaw point.

ating simultaneously in the Fairbanks area during the summer of 1929. Some 200 men (typically called "point doctors") hammered and twisted these points into the ground and maintained the flow of water through them. Complete thawing of the 10- to 20-meter thick frozen paying gravels required 45 to 100 days, depending on ground conditions and point spacings. This was a tremendous and costly undertaking because each of the several dredges operating in the Fairbanks area processed 2 to 5 million cubic meters of material each season, and most of that had to be thawed. Estimates at the time indicated that thawing the permafrost in the gravels and the thawing and hydraulic removal of the frozen overburden accounted for half the overall cost of mining gold with dredges. Modern-day gold miners no longer use the huge arrays of cold-water thawing points, but they continue to use dredges and hydraulic giants, and often make use of solar heat to thaw the ground. When exposed to

Figure 5.7 A field of cold-water thaw points.

the sun, frozen ground might thaw ten centimeters or more in one summer day in interior Alaska. Because the rate of thaw declines rapidly with depth, most modern miners there use caterpillar tractors, excavators and trucks to remove material as it thaws (or even before it thaws), in part because this technique is more environmentally acceptable than the use of water to transport the overburden.

■ Building on Permafrost

The sod igloos of bygone times were forgiving of any settling from thawing of ice-rich permafrost below. Log cabins built directly on the ground also may continue to be almost as livable as they ever were as they sink into the thawing ground, and sometimes the rotting of the bottom logs

Color Plates

Plate 1. An island covered by dying trees sinks into a thermokarst lake near where the Alaska Highway crosses the Alaska-Canada border, in a region where warming during recent decades has caused shores to collapse as the permafrost below melts. See also Figure 1.4, page 8.

Plate 2. A forest of refrigerated pilings supports the trans-Alaska pipeline where it snakes through the Alaska Range north of Isabel Pass. See also Figure 1.7, page 12.

Plate 3. Pipkrakes formed during several cold nights at sea level on San Juan Island, Washington. The arrows mark the lower edge of the pipkrake layers. The two photos at bottom show pipkrakes removed from the location shown in the photograph at upper right. See also Figure 3.1A, page 46.

Plate 4. Is this the grandaddy of all pipkrakes? Billy Conner, a researcher with the Alaska Department of Public Facilities and Transportation, is shown with a substantial specimen. Perhaps formed by another mechanism, this large ice mass protruding approximately 1 meter above its surroundings was photographed by Richard Veazey near the trans-Alaska pipeline on Alaska's North Slope. See also Figure 3.1C, page 48.

Plate 5. Soil uplifted by pipkrakes away from pebbles in a wet driveway on San Juan Island, Washington. See also Figure 3.2, page 50.

Plate 6. Ice wedges and multiple layers of segregation ice in a roadcut near Fox, Alaska. See also Figure 3.19, page 92.

Plate 7. A layer of pure ice pushing up the moss in northern Alaska. Photograph by Richard Veazey. See also Figure 3.22, page 97.

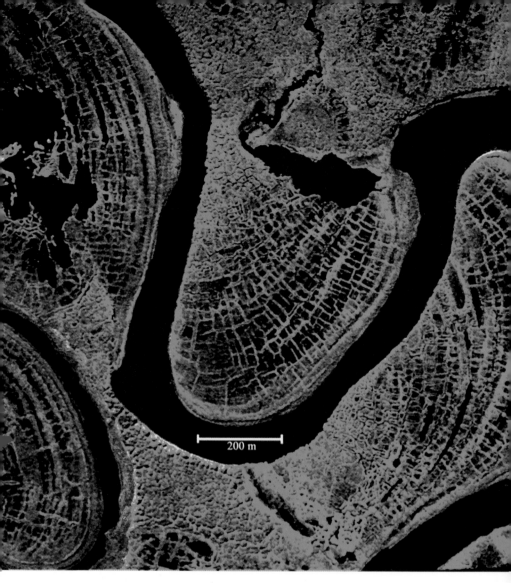

Plate 8. High-altitude infrared photograph (false color) of ice-wedge polygons on Alaska's North Slope. Courtesy Geophysical Institute, Fairbanks, Alaska. See also Figure 4.4, page 108.

FOXY

65 cm

Plate 9. Top: Photo of thermokarst topography (high-center polygons) in an abandoned field taken in 1996, two days after the aerial photograph shown below. Microrelief in the thermokarst area is near 1 meter. On the higher ground such severe cracks parallel the depressions that it would be unwise to ride a horse across this field, located near the eastern end of the Farmers Loop permafrost area near Fairbanks. The scale dog seen in this and other illustrations is Foxy, a sheltie-husky cross 1 m long from tip of nose to base of tail. She stands 50 cm tall at the shoulder, and when sitting the top of her head is about 65 cm above ground. See also Figure 4.6A, page 112.

Plate 10. Top: The collapsed open-system pingo near Fairbanks, Alaska, shown in Figure 4.20, page 130. Bottom: Close-up view of two of the mud cones formed by influx of mud-charged underground water flowing into the pingo.

Plate 11. A symmetrical palsa in the complex Palsas at Mile 206.7 on the Parks Highway just south of Cantwell, Alaska. See also Figure 4.22C, page 135.

Plate 12. An elongated palsa in the background with one that is collapsing in the foreground. Photographed fall 1995. See also Figure 4.22D, page 136.

Plate 13. Top: A turf (bog) hummock; bottom: an earth hummock showing cryoturbated soil layers, both photographed along the Denali Highway between Miles 30 and 37 in 1997. See also Figure 4.23, page 138.

Plate 14A. Top: A small rock glacier photographed from the Richardson Highway at Mile 208 in 1996. The white dots mark the approximate upper boundary of the glacier, and the white arrow marks the location of the photograph at bottom and in Plate 14B. Bottom: A view directly across the rock glacier's frontal ramp toward the less steep collovial slopes beyond.

Plate 14B. Top: The 1-meter scale dog Foxy scrambles directly up the frontal ramp; and at bottom, slinks unhappily across it after discovering that every footstep can start a slide of the loose rock and soil lying at its angle of repose. See also Figures 4.33A and B, pages 162–163.

Plate 15. Sorted polygons at Mile 31.8 on the Denali Highway, Alaska. Top: Lightly vegetated soil showing thermal or desiccation cracks at lower-right corner grades into the sorted polygons at center, and these grade into the rock pavement beyond, the sequence illustrating the transition with time. The close-up views at bottom show newly exposed white rocks in the central part of polygons. (The two shown in the right-hand picture are the two nearest the camera in the left-hand picture.) Notice that the rocks grow darker the longer they are exposed, and that the boundaries between the inner fines and the outer rocks are trending toward circular. Photographed August 1996. See also Figure 4.39, page 177.

TREAD

RISER

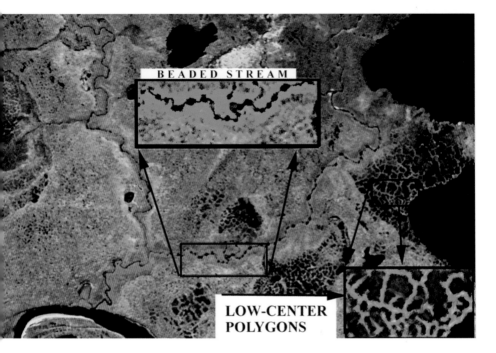

Plate 17. Beaded stream drainage and low-center polygons in a high-altitude infrared photograph of Alaska's North Slope. See also Figure 4.44, page 189.

te 16. Cryoturbation steps
oroximately 10 km north of
che Creek, B.C., on Highway 97.
p: view along the cryoturbation
p identified by the arrow in the
ddle photograph (taken
ultaneously by Rosemarie
vis), and at bottom, a more
tant view of the array, looking
th. Photographed in 1996. See
o Figure 4.43A, page 183.

Plate 18. Two views of thaw lakes bordered by tilting and submerging trees, signaling that the lakes are undergoing rapid enlargement. The upper photograph was taken in September 1996 at Mile 1125.5 on the Alaska Highway and the lower along the highway some miles to the north, also in September 1996. See also Figure 4.45, page 190.

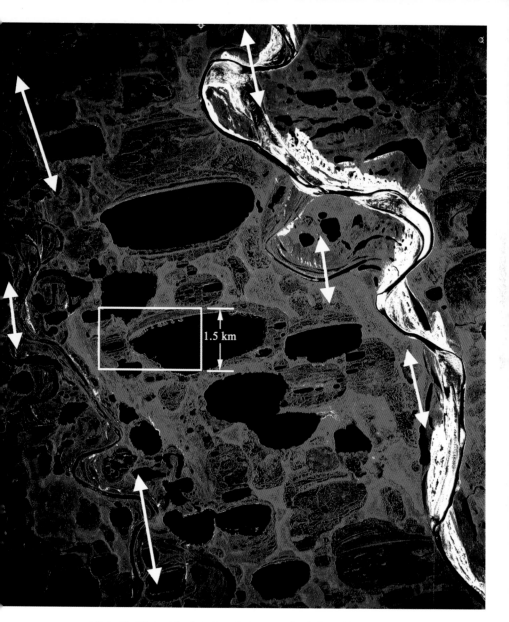

Plate 19. High-altitude infrared photograph of aligned lakes on Alaska's North Slope near Point Barrow. White arrows indicate approximate alignments of nearby snowdrifts, seen in the photograph. The portion within the white box is enlarged as Plate 20. See also Figure 4.47, page 193.

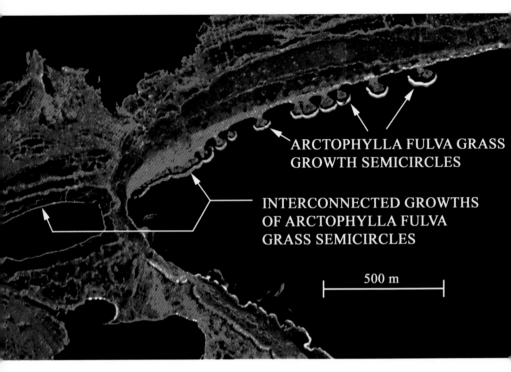

ARCTOPHYLLA FULVA GRASS
GROWTH SEMICIRCLES

INTERCONNECTED GROWTHS
OF ARCTOPHYLLA FULVA
GRASS SEMICIRCLES

500 m

Plate 20. An enlarged view of the area shown in Plate 19. Around the small lake at left former shorelines are evident, each apparently created by joined semicircular growths of vegetation. See also Figure 4.48, page 194.

Plate 21. An Eskimo summer camp with facilities placed on the driest places, the middle parts of high-center polygons (see arrows). The trail is made by dog sleds run across the tundra in summer. Photographed near Point Hope, Alaska, in 1957. See also Figure 5.1, page 202.

Plate 22. The remains of an early-day mining shaft west of Ester, Alaska. See also Figure 5.2, page 203.

Plate 23. Reconstruction underway on the foundation of a house adjacent to the University of Alaska campus in a subdivision where many of the homes have settled unevenly. The house is shown resting on steel beams after having been jacked up above its partially collapsed concrete foundation walls. Subsequent to these photographs taken in summer 1996, the owner-builder removed the steel beams and set the house back down on a thick concrete foundation placed atop 750 cubic yards of gravel imported to the site. He stated that he believed the collapse of the old basement was due to settling of unstable water-saturated soil rather than the melting of segregation or wedge ice. See also Figure 5.16, page 223.

ADJUSTMENT SCREW

Plate 24. New three-story log home constructed by owner Matthew Carlin, shown at bottom. He is fully aware of the site's underlying permafrost at depth near 6 m and has planned well to deal with it. The foundation beneath this heavy structure (which includes a multicar garage with insulated concrete slab floor) rests atop a formidable array of interlocked steel beams resting on large-diameter adjusting screws. Part of the project involves moving the adjacent log cabin (which over the years had settled 1 meter on one end) to a new site where it too is being placed on an adjustable steel foundation. See also Figure 5.18, page 225.

contributes to the settling. Some log cabins, with or without foundations of concrete or other materials, have survived for many years while others rapidly failed. Concrete or wooden frame buildings on concrete foundations are particularly susceptible to serious damage when ice-rich permafrost below them melts.

If the permafrost beneath a structure built in contact with the ground contains evenly distributed ice, the ground may subside so uniformly when the ice melts that the building is not distorted. However, melting of unevenly distributed ground ice or uneven thawing, such as might be caused by the placement of heating plants, can seriously distort buildings, as is shown in **Figure 5.8A** and **B**. Buildings constructed over ice wedges or atop pingos (see Figures 1.8 and **5.9**) invariably have serious problems unless special precautions are taken to avoid thawing the ground below.

Why do people build houses atop pingos? Because the top of an open-system pingo can be appealing: rising above the usual swampy black spruce forests below, the pingo typically sports an open canopy of mixed coniferous and deciduous trees that creates a pleasant park-like setting for a home, and it is one with a view. Some home builders perhaps do not recognize that the little rises are pingos nor do they understand the likely consequences of building atop them. Others are aware but go ahead anyway, building their homes on piling foundations that can be adjusted and that help prevent heat flow from the house to the ground. This was the approach taken by the builders of the several homes on the pingo complex shown in Figure 4.19.

On the north side of Fairbanks, just east of the University of Alaska campus, is an area several miles long that has experienced many problems with permafrost over the years. Thanks mainly to long-term study by the well-known geocryologist Troy L. Péwé, much is known about this area and its problems. As shown by his illustration reproduced as **Figure 5.10**, an extensive layer of permafrost underlies a broad alluvial apron and extends up some distance along the lower slopes of the hills. The permafrost is up to 80 m thick and is mainly within fine-grained silty loess transported downslope chiefly by runoff water from the melting of ice-rich permafrost plus lesser transport by solifluction. The silty soil is conducive to cryosuction, so it is not surprising that this permafrost is rich with massive layers and fine seams of ice distributed through the upper permafrost layer. A number of wells drilled along the northern boundary of the area have yielded artesian flow, indicating that much of the layer is under several atmospheres of hydrostatic pressure, because water coming down off the hills to the north is trapped beneath the permafrost layer. Thus the

Figure 5.8A Leaning together for mutual support, these two frame buildings built during the Klondike gold rush era show evidence of melting permafrost beneath their foundations, as do many other nearby buildings in Dawson City, Yukon Territory. Heather Star Davis stands on the boardwalk. Photographed in 1998.

Figure 5.8B An early-day Dawson store with a foundation problem, photographed by Bon V. Davis, probably in 1952.

Figure 5.9 Unable to withstand approximately 1 meter of differential settling, this poorly constructed log cabin built atop an open-system pingo beside the Elliott Highway northwest of Fairbanks, Alaska, was abandoned, but the current occupant in 1996 was accumulating timbers (in foreground) in preparation for reconstruction.

water supply to the permafrost layer is not only ample, it is there under a pressure that can augment cryosuction in forming ice within the permafrost layer. Since the temperature near the top of the permafrost in the Fairbanks area is right at the melting point, any disturbance that increases

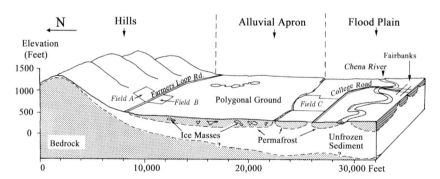

Figure 5.10 Diagrammatic sketch showing the character and distribution of permafrost in the Fairbanks, Alaska, area. Field A is not underlain by permafrost. Field B is underlain by permafrost containing large ice masses; thermokarst mounds and pits may form when the ice melts. Field C is underlain by permafrost without large ice masses; mounds and pits will not form at the surface and ground subsidence is negligible. Reproduction of Figure 3 of Péwé (1982).

soil temperature will cause some melting. The same is true near the bottom of the permafrost layer—the temperature is at the melting point—so the flow of water in contact with it may cause some melting.

One spectacular consequence of the artesian situation in parts of this area is that a well drilled near the east end of Farmers Loop Road in 1976 flowed out of control. Enough permafrost melted near the well casing that attempts to cap the well resulted in a copious flow of artesian water up around the casing, a flow that continued well into the winter and froze into a thick layer of ice that blocked traffic on Farmers Loop Road. Before it froze, the ice flowed into houses and around automobiles. According to one report, it was possible to walk on the ice and look down at the top of a Volkswagen sedan just below the ice surface.

At the south boundary of the Farmers Loop permafrost area is a sharp transition between deposits of ice-rich silt and frozen sand and gravel that has low ice content. **Figure 5.11** illustrates the relevance of this boundary to local land owners. Buildings constructed on the coarse-grained material to the south of the boundary have stood up well over the years, while the tenure of those on the north, especially if constructed with concrete foundations, has been short.

Travelers along Farmers Loop Road for several years had the chance to watch the three components of the house shown in **Figure 5.12** jostle each other into submission as they tilted and sank into the ground. All that remained by 1995 was the shattered concrete foundation. Several other houses in the nearby area show signs of serious distortion, as do several segments of Farmers Loop Road itself only two years after an expensive reconstruction.

One of the easily recognized signs of ice-rich melting beneath structures is the raised porch syndrome depicted in **Figure 5.13**. Often misinterpreted as caused primarily by upward annual frost-heaving of the coldest part of a structure, the elevated porch or garage is merely that part of a structure that has settled the least over the course of some years.

The approaches people take to encountered or expected problems with housing built on permafrost cover a broad spectrum. At one end is denial that a problem actually exists or that melting ice-rich permafrost causes it. When Farmers Loop Road was reconstructed in 1994 and 1995 some residents blamed the sinking of their houses on the vibrations caused by heavy construction equipment moving along the roadway. They may have been partially correct to the extent that these vibrations perhaps caused enough settling of saturated silty soil along the roadway to affect nearby house foundations. Certainly anyone standing nearby could *feel* the passage of the heavy machines as they rumbled past.

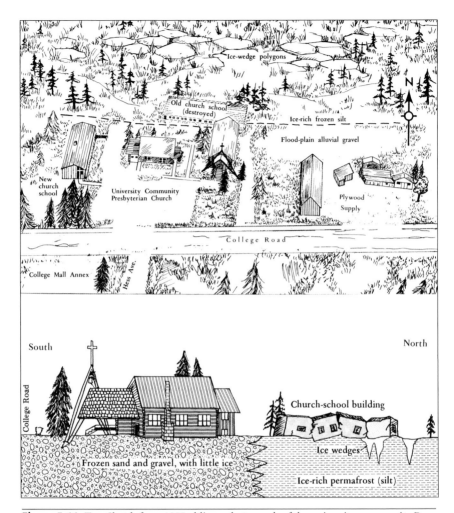

Figure 5.11 Top: Sketch from 1955 oblique photograph of the university community Presbyterian church manse and old church school in College, near the southeast corner of the University of Alaska Fairbanks campus. Buildings in foreground are on flood-plain alluvial gravel and church school is on ice-rich permafrost. Note the ice wedges diagrammatically expressed as polygonal ground. The old school collapsed several years later as underlying ice melted. The new school building and other modern structures are sketched as of 1981. Photography courtesy of E. H. Beistline. Bottom: Diagrammatic sketch of sanctuary of the church and former school building. The sanctuary was built on perennially frozen flood-plain sand and gravel that contains scant ice. Thus, the building has subsided little. In 1955, the former church school building was constructed 30 feet to the north, on ice-rich permafrost, across the boundary between the flood-plain gravel and retransported silt. Permafrost under the heated block building thawed, and as the building subsided it was distorted and had to be razed 6 years later. Reproduced from Figures 49 and 50 of Péwé, 1982.

Figure 5.12 Drawn-in lines emphasize the differential settling that has occurred as this house beside Farmers Loop Road over the course of several years. It was abandoned in 1993 or 1994 when the top photograph was taken from the site of the automobile in the lower photograph. Taken in 1995, it shows all that is left, a highly distorted concrete pad.

HEATED PORTION SINKS MOST

Figure 5.13 Houses suffering the raised-porch syndrome, actually caused by preferential sinking of the warmest parts of a structure, as indicated in the top photograph of Bert and Mary's Roadhouse near Big Delta, Alaska. Built in 1949, the main structure began sinking in 1952 and subsided approximately 60 cm by fall 1953. The building was still in use by 1959 but was abandoned by 1962 when Troy L. Péwé took this photograph (Péwé, 1982). The middle photograph taken in 1997 shows the breakage and sinking of a concrete block basement beneath a prefabricated home located in Goldstream Valley north of Fairbanks. At bottom is a cabin built of railroad ties that held together well as it sank approximately 1 meter over the course of several decades. It is located on Farmers Loop Road just east of Field B in Figure 5.9. Photograph taken July 1996.

In evaluating the role of melting ice-rich permafrost in the settling of houses—as contrasted to distortions to foundations possibly due to other causes—people probably often underestimate how effectively a heated house can melt frozen ground located many meters below its foundation. Test drilling that shows no permafrost within a few meters of the ground surface can easily lull a home builder into thinking that no problems with permafrost will be encountered. Yet there are several, and perhaps many, homes in the Farmers Loop and other permafrost areas that have experienced damaging settling within a few years even when test drilling encountered no permafrost within 10 m of the ground surface.

One well-documented example involves two adjacent homes located about 100 m south of Farmers Loop Road some 5 km from its eastern end, and not far from the house with the raised-porch syndrome shown at bottom in Figure 5.13. At the time the one single-story house with a full concrete block basement was built in 1966, the drilling of a water well showed permafrost at depth approximately 11 m. Over the next 15 years the slightly heated (to about 4.5°C) basement underwent settling, mostly toward one end, and some inward sloping of the concrete block walls developed. Then in 1981 the owner insulated the basement and began maintaining its temperature near 20°C. Settling then accelerated rapidly during the next 3 years and became so serious that the house was jacked up and then reestablished on the site after filling the basement with dirt. The estimated cost of the repair was $25,000 plus the loss of the basement's usefulness. In 1984, test drilling near the house showed the top 11 m of the soil to be unfrozen clayey silt.[3] Just 15 cm below the top of the permafrost the test drilling encountered a layer of pure ice nearly 1 m thick, and below that two other ice layers 15 cm thick. Down to where the drilling stopped, at depth 16 m, measurements showed the volume of ice in the permafrost was near 40%.

Next door to the east sits the house shown in **Figure 5.14**, one constructed in the mid-1980s with a full poured-concrete basement that was kept heated at room temperature at the outset. Serious settling took place during the first few years, necessitating the placement of up to 30 cm of steel beam and wood shims to level the floor above and a major effort to stop the inward collapse of the basement walls. The effort includes the placement of a steel beam across the center of the basement midway up the walls.

---

3. Abbott (1984).

Figure 5.14 This heated-basement residence at 3 Mile Farmers Loop Road began sinking almost as soon as it was built, and by 10 years later Glenn Bouton, the owner, had taken elaborate steps to correct for the subsidence and prevent further buckling of the basement's concrete walls. These included the insertion of shims as shown in the upper and lower right photographs and a series of interconnected and adjustable steel truss supports on the inside and outside of the west wall of the basement. Photographed in summer 1996.

It might seem the enclosed foundations of these two houses initially were high enough above the top of the permafrost that they would not induce melting, but rough calculations of the rate of melting expected from heat introduced to the ground by the houses indicate otherwise. These calculations make use of a simple equation (the Stefan equation)[4] that allows determination of how much frozen ground can be thawed by a given amount of energy input if enough is known about the thermal characteristics of the soil involved. In general that information is missing, as it is here, so the results depend on several assumptions combined with observations of Fairbanks area temperatures and of the rates that energy moves down through the ground.

During an average year, Fairbanks has approximately 1800 degree-thawing-days (number of days times average temperature in °C above 0°C), and this annual warm spell causes a wave of increasing temperature to penetrate downward from the ground surface at a rate that depends on the thermal conductivity of the soil, its water content and whether it is frozen. The energy entering the ground first melts all ice in the active layer, which at Fairbanks has thickness less than 1 m in mossy areas and which may exceed 2 or even 3 m at locations kept clear of vegetation. Where the top of the permafrost is below the bottom of the active layer, the summer temperature increase then warms the unfrozen soil down to a depth typically of 8 to 15 m, a range that brackets the 11-m depth to permafrost beneath the two houses in question.[5]

Figures 5.15A and **5.15B** show the results of my calculations using the Stefan equation cited above, based on Fairbanks weather records, temperature, and soil composition data from the test hole drilled near one of the houses in 1984, and several assumptions: 1) the annual summer temperature

---

4. The form of the Stefan equation adopted here relates the square of the depth thawed or elevated in temperature by a given amount to a quantity proportional to (energy available) (thermal conductivity)/(energy required to melt frozen ground or raise its temperature a given amount). To compile Tables 5.1 and 5.2 the adopted form is D^2 (meters) = K(Thawing Degree-Years), where K = 0.041 in the unfrozen case and K = 0.437 in the frozen case.

5. In general, the maximum temperatures produced by the summer pulse occur during midsummer near the top of the ground, but progressively later with depth so that at depth 3 m the maximum temperature occurs in September or October, and at depth near 10 m the maximum temperature is likely to occur as late as the following January. The information in plots of temperature versus time made to represent various depths permits a determination of how rapidly energy moves through unfrozen soil.

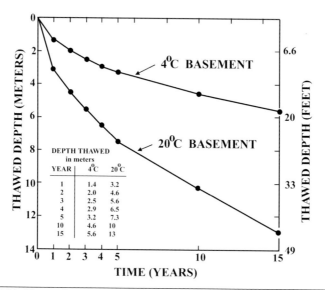

DEPTH THAWED in meters		
YEAR	4^0C	20^0C
1	1.4	3.2
2	2.0	4.6
3	2.5	5.6
4	2.9	6.5
5	3.2	7.3
10	4.6	10
15	5.6	13

Figure 5.15A Plots and numerical results showing the expected progressive depth thickness in meters (feet) of frozen ground thawed by houses with basement temperatures 4°C and 20°C if set in contact with frozen ground, given the assumptions stated in the text.

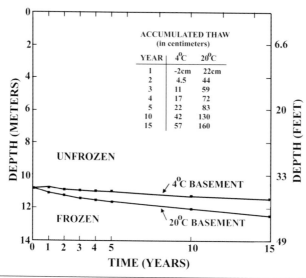

ACCUMULATED THAW (in centimeters)		
YEAR	4^0C	20^0C
1	-2cm	22cm
2	4.5	44
3	11	59
4	17	72
5	22	83
10	42	130
15	57	160

Figure 5.15B Plots and numerical results showing the expected accumulated thawing of permafrost caused by houses with basements maintained at 4°C and 20°C placed on unfrozen ground over permafrost at initial depth 11 m, and assuming that the house replaces the ambient summer heating during the first year. See text for other assumptions.

pulse generates a change in temperature near 1°C at a depth above but close to the top of the permafrost (11 m); 2) the composition of the soil column is uniform throughout and everywhere contains 25% water by weight in both frozen and unfrozen portions (based in part on the well data); and 3) the energy delivered to the top of the permafrost is all available for melting ice. Assumption (1) is crucial because it fixes the rate at which heat is delivered to the top of the permafrost. Identical results are obtained if this assumption is replaced by another: that the summer heat pulse at Fairbanks is able to melt 1.6 m of frozen ground having the composition and water content of that beneath the two houses.[6] The results also are in agreement with the result of a theoretical calculation described by Lachenbruch (1957). The agreement among the three approaches is pleasing; nevertheless, the results shown in Figure 5.15A and 5.15B should be viewed as only qualitative at best. For one thing, the soil composition and water content are not likely to be uniform throughout the column. Also, this treatment is one-dimensional; that is, it holds strictly only when the area of the building involved is large compared to the depth considered (see Lachenbruch, 1957.)

Figure 5.15A shows—assuming correctness of assumptions made— the approximate thickness of frozen ground that should be thawed by a house with basement temperature 4°C set directly atop frozen ground, and one similarly placed but with the basement kept at room temperature (20°C). Notice that the rate of thawing declines with the passage of the years, and that the house with the warmer basement thaws the frozen ground far more effectively than the one with the cold basement. The thaw basin beneath the warm house is approximately 7 m deep after five years whereas that beneath the cooler basement is less than half as deep.

Of greater interest are the results in Figure 5.15B indicating the time taken for the temperature wave generated by each house to penetrate unfrozen ground. Unlike a seasonal pulse, the heat flow, while initially transient, sets up a permanent temperature gradient that feeds energy to the top of the permafrost. If no permafrost is below a house, the rise in

---

6. Measurements of annual freeze and thaw depths in undisturbed silty soils in the Fairbanks area show much variation from year to year, from less than 1 m to more than 2 m [Lotspiech (1973)]. An independent calculation assuming Fairbanks area temperatures and a uniform silty soil with 15% water content and no snow cover yields annual freeze and thaw depth 2.1 m; the method follows procedures described in a technical manual for military engineers [Army (1966)].

temperature sweeps downward rapidly, and radially outward as well.[7] The diagram indicates the amount of permafrost that might be melted in the case of the two houses on Farmers Loop where the top of the permafrost is assumed to lie initially at depth 11 m and it is assumed that the heat energy delivered to that depth goes into melting the permafrost instead of propagating the temperature increase downward. Although the actual numbers shown probably are not strictly correct, they indicate that a few centimeters of permafrost will melt during the first year. Because the drill hole showed a thick layer of pure ice just below the top of the permafrost, it is quite understandable that the houses both began to sink during the first years after construction. The potential for melting beneath a house with a basement maintained at room temperature year around appears to be especially large, in excess of 10 cm per year (or in excess of 5 cm of a layer of pure ice)—even when the permafrost is 10 m or more below the house's foundation.

Although the numerical results contained in Figures 5.15A and B are best treated as only approximate, they and actual experiences with houses collectively convey a very important message to those owning or wishing to build houses on ground that might contain permafrost: Even if the ground below the active layer is unfrozen, danger may still lurk below in the form of layers of segregation ice in fine-grained permafrost perhaps combined with an overlying stratum of saturated soil that may have low bearing strength. In this situation, the top portion of the permafrost is right at the melting point, so just the slightest increase in energy fed downward to it will cause melting, and this top part may well contain extensive lenses of segregation ice or ice wedges, perhaps both.

In some instances where the permafrost is warm enough that it is unlikely to form again once melted, builders have chosen to thaw the frozen ground before placing buildings on it. One example involved the construction of a large post office building on discontinuous permafrost near Fairbanks (Alaska) International Airport in 1978. Engineers used steam points driven into 315 holes to thaw a body of permafrost down to a depth of 12 m. The project took several months and thawed 26,000 cubic meters of soil at an expense of approximately $7 per cubic meter. Twelve years later, the post office building exhibited some cracking and subsidence.[8]

---

7. At depths greater than the shortest dimension of the house or other heated structure, the ground temperature is affected by outside air temperature as well, so actual temperature increase at depth will be less than suggested by the table entries.
8. McFadden and Bennett (1991) pp 26–30.

Increasingly in the Fairbanks area, elevated steel foundations are coming into use for new homes built on ice-rich permafrost and structures needing foundation repairs because of melting permafrost. Consisting of steel I-beams supported by adjustable steel piers, these foundation structures reportedly cost roughly the same as a concrete slab foundation. See **Figures 5.16** to **5.18**. Disadvantages of an elevated foundation with a cold air space below are that floors need to be heavily insulated and special precautions are necessary to prevent above-ground freezing of water and sewer lines.

Using such adjustable steel foundations is but one approach to the permafrost problem. In locations where the mean annual temperature is well below 0°C, it may be possible to maintain a stable foundation by placing the buildings on pilings that permit circulation of enough ambient air

Figure 5.16 Reconstruction underway on the foundation of a house adjacent to the University of Alaska campus in a subdivision where many of the homes have settled unevenly. The house is shown resting on steel beams after having been jacked up above its partially collapsed concrete foundation walls. Subsequent to these photographs taken in summer 1996, the owner-builder removed the steel beams and set the house back down on a thick concrete foundation placed atop 750 cubic yards of gravel imported to the site. He stated that he believed the collapse of the old basement was due to settling of unstable water-saturated soil rather than the melting of segregation or wedge ice. See also Plate 23.

ADJUSTMENT SCREW

Figure 5.17 This house, located in an area of raised-center polygons on a hillside north of the University of Alaska campus in the College Hills subdivision, underwent settling and then was jacked up and placed on steel I-beams resting on adjustable jack posts. Notice the insulated sewer and water utility pipes, a necessary feature of homes placed on piling foundations.

ADJUSTMENT SCREW

Figure 5.18 New three-story log home constructed by owner Matthew Carlin, shown at bottom. He is fully aware of the site's underlying permafrost at depth near 6 m and has planned well to deal with it. The foundation beneath this heavy structure (which includes a multicar garage with insulated concrete slab floor) rests atop a formidable array of inter-locked steel beams resting on large-diameter adjusting screws. Part of the project involves moving the adjacent log cabin (which over the years had settled 1 meter on one end) to a new site where it too is being placed on an adjustable steel foundation. See also Plate 24.

to keep the foundation frozen or by inserting large air ducts in the foundation for the same purpose. These techniques have proven generally successful for stabilizing the foundations of hangars and other large structures built in the early 1950s at Thule, Greenland, where ice makes up as much as 50% of the soil's volume. In the few instances where ducts did not allow adequate air flow the buildings at Thule underwent settling, one large hanger by 3 m at its center.[9]

Among the most impressive of structures ever constructed atop permafrost are two huge buildings composing the main part of the BP Alaska/Sohio North Slope Operations Center in the Prudhoe Bay oil field. According to Peter Floyd, the well-known architect who designed the buildings, the key is to acknowledge the peculiar problems and to analyze them carefully, and in this it seems he was successful. The larger of the two structures consists of four prefabricated modules each three stories high, 15 m (50 ft) wide, 38 m (126 ft) long, and weighing 750 tons. To avoid accumulating blowing snow and to keep the ground below frozen, the building stands on pilings that extend 2 m above a thick gravel pad placed atop the ground. Set into predrilled 10 m-deep holes (drilled during winter to avoid thawing), the huge steel support pilings, up to 70 cm in diameter, contain insulating elements designed to minimize heat flow into the ground. Since a building of this size undergoes substantial expansion and contraction during changes in temperature, the pilings also incorporate joints along which horizontal slippage can occur.[10]

Another technique for preserving frozen ground beneath buildings and other structures is the installation of passive heat pumps (also called *thermopiles*, thermotubes, convection tubes, and thermosyphons). These can provide a one-way transfer of heat from the ground to the air whenever the air is below a certain temperature. In its simplest form such a heat pump consists merely of a sealed pipe charged with a fluid such as glycol or alcohol and water. It relies on the working fluid's change in density with temperature to generate convective flow that removes heat from the ground. Usually the pipe is set vertically and is equipped at top and bottom with external fins to promote the heat exchange, the upper fins aboveground and the lower below. Another type of heat pump depends on changes between liquid and gas phases of Freon, butane, propane, or anhydrous ammonia working fluids to effect the heat transfer. Ammonia, for example, boils at –33°C if held at sea level pressure (1 Atm), but the boiling

---

9. Mangus (1986).
10. Floyd (1974).

point rises rapidly with pressure until at a pressure of 3 Atm the boiling point is approximately 20°C. Therefore it is possible to set the boiling point temperature in the tube merely by adjusting the charging pressure. If the charging pressure is adjusted to approximately 2 Atm the boiling point will be –2°C. Then if the ground and air temperature are both above –2°C the pipe will contain only ammonia in vapor form, and no heat exchange will occur (or if the ground temperature is below –2°C the pipe will contain liquid ammonia at its bottom, but still no heat exchange will occur). However, if the air temperature falls below –2°C and the ground temperature is above –2°C, the ammonia vapor will condense to liquid on the upper walls of the tube, run down the walls and revaporize near the bottom. Because the heat of vaporization of ammonia is large, approximately 330 calories per gram, and its heat capacity is about equal to that of water, the vaporizing ammonia is able to withdraw substantial heat energy from the ground in winter and transport it to the air above. Thus, if properly pressurized, the ammonia heat pump can keep ground frozen, or cause thawed ground to freeze. The finned objects shown in Figure 1.7 are a few of the 120,000 ammonia-charged heat pumps that keep the ground frozen around the vertical support members of the 390 miles of elevated portions of the trans-Alaska oil pipeline.[11]

Figure 5.19 shows a reasonably successful application of the passive heat pumps to stabilize the foundation below a major structure built near the Steese Highway north of Fairbanks in 1984 by the OK Lumber Company to house a truss assembly plant. Preliminary drilling revealed that the site—located near the northeast boundary of the extensive Farmers Loop Road permafrost area—contained frozen ground at depth 6 m. Prior to construction of the two metal buildings, a thick layer of dredge tailings was laid down, and the overall site was leveled. Expecting problems if the frozen ground thawed, the owners invested more than $100,000 in an extensive array of passive heat pumps. The buildings (one heated, one not) were used for their intended purpose for six years; however, during the first year some sinking occurred below the central part of the heated building. Yet by 1996, when the photographs in Figure 5.19 were taken, the two buildings appeared to be plumb even though the surrounding yard area showed severe settling in a polygonal pattern, indicating melting ice wedges below.

During the 1970s, the construction of a telemetry site at the University of Alaska's Poker Flat Research Range north of Fairbanks accidentally

---

11. McFadden and Bennett (1991) pp 46–47, 195–202.

PASSIVE HEAT PUMPS

Figure 5.19 Except for a problem in the central area of one building, heat pumps installed to prevent thawing beneath the two were successful in keeping the buildings level. Thermokarst topography due to thawing ice wedges developed in the yard area (foreground of top photograph). Photographed in 1996.

triggered the formation of a small open-system pingo. A well drilled at the site reportedly penetrated several tens of meters of pure ice overlain by a sticky clay that was also exposed during clearing operations. Realizing that they had uncovered a serious problem, the operators immediately graded material back onto the site and installed passive heat pumps in an attempt to keep the ice from thawing. The effort was successful enough to allow use of the site but a small pingo began to grow near one edge of the area initially cleared.[12]

■ Poles, Pilings, and Permafrost

Posts placed in frozen ground to support such things as fences, communication lines, power lines, buildings, and bridges are subject to two kinds of problems: the frost heaving discussed in Chapter 3, and settling caused by the melting of ice-rich soil. In attempting to construct an overland telegraph line that would link America and Europe via Bering Strait during the 1860s, the Western Union Telegraph Company avoided both problems by setting above-ground tripods to support the telegraph wires where they crossed northern British Columbia and Alaska. This method also avoided digging holes in frozen ground, and it continued to be used well into the next century. Tripods and guyed bipods also have been used successfully to support power transmission lines in tundra areas of Alaska—somewhat to the chagrin of linemen who quickly discovered that climbing a bipod or tripod to make repairs was not nearly as easy as climbing a vertical pole. Another generally successful approach to both problems is to set vertical poles in rock-filled cribs, but these are too costly in labor and materials to be used except in special situations.

The main electrical power distributor in interior Alaska, Golden Valley Electric Association, is experimenting with the use of passive thermal piles to keep the ground frozen around power poles set in ice-rich permafrost. **Figure 5.20** shows one installation at a location just uphill from the pingo identified as "B" in Figure 4.19. Here, the permafrost table is shallow and therefore subject to thawing even with only minor disturbance to the surface cover. Along this stretch of power line, just the clearing of brush and trees has allowed enough thawing to create thermokarst topography everywhere on the right-of-way.

---

12. Neal Brown, personal communication, 1998.

Figure 5.20 A passive heat pump intended to prevent settling of a power pole located just above pingo B shown in Figure 4.19. Photographed in 1996.

The degree of difficulty encountered in placing posts or pilings in frozen ground depends much on the size of the soil particles involved. Wood and concrete pilings cannot be driven into frozen ground, but it is possible to drive open-ended steel pipe and H-shaped steel pilings into fine-grained and peaty permafrost. **Figure 5.21** illustrates that ice-cemented gravel is not

Figure 5.21 Top: A circular steel piling crumpled by an attempt to drive it into frozen ground. Bottom: An I-beam piling that suffered the same fate. Both are on display in the halls of the Duckering (engineering) Building at the University of Alaska in Fairbanks.

nearly as amenable to accepting similar penetrations.[13] One method is to thaw frozen ground with a heat source prior to driving pilings, and another is to predrill holes to accept the pilings. Rotary drills with teeth designed to

---

13. Rice (1973).

shave off thin slices of material are able to penetrate frozen clay, silt, and peat soils, but drilling into coarse gravels requires teeth that project out from the drill face so that they tear rather than shave. Drill bits wear rapidly in frozen gravel, and if the material is extra coarse the drilling is difficult.[14]

An advantage to predrilling is that the pilings can bear weight as soon as placed in frozen ground. Backfilling should be done with dry material to prevent thawing, and if the ground is thawed to place the pilings then they should not bear loads until the ground is completely frozen.[15] Evidently, numerous foundations have failed quickly because of insufficient time allowed for refreezing. Furthermore, pilings with their lower parts well gripped by permanently frozen ground will not be subject to frost heaving when soil in the active layer freezes and thaws. Engineer Eb Rice of the University of Alaska Fairbanks, a well-known authority on such matters, noted that for this reason wooden pilings should be placed with the big end down. Wrapping the upper portions of piles with plastic or some other slick material also can reduce frost jacking by reducing the grip of freezing soil in the active layer.

▪ Water Supply and Sewage Disposal in Permafrost Areas

The problems of water supply and sewage disposal (also called wastewater disposal) in areas of continuous and discontinuous permafrost link together for several reasons. Among them are: 1) frozen ground is highly impermeable to both, 2) conventional piping of both water and sewage is difficult when the ground is frozen, 3) frozen ground tends to limit the availability of water, and that influences the makeup and amount of sewage requiring disposal, and 4) thawing of ground beneath water supply and sewage treatment facilities can lead to failures in these facilities. The approaches to these problems depend much on the size of the social entity involved. Individual families living in remote locations typically solve the problems simply and at low or zero cost by using hand-carried surface water and outhouses, and people living in small villages may do the same. At the other extreme are elaborate distribution, collection, and treatment systems so expensive and complicated to build and operate in the north

---

14. Reimers (1980).
15. The vertical support members used to support the elevated portions of the trans-Alaska oil pipeline were placed in winter and backfilled with a sand slurry that then froze.

that only moderate to large cities or regional or federal organizations can provide them—and not always very well. In general, government is prone to undertake new utility projects in permafrost areas without allocating sufficient attention and funding to long-term maintenance.

Water in areas of continuous permafrost

Convenient sources of water may be difficult to find in permafrost regions, and the water may be of poor quality. Where permafrost is continuous and thick, surface water or water contained in thawed strata below lakes and rivers usually are the only viable sources. (Another source in arctic areas is melted ice, either freshwater ice or sea ice made salt-free by aging over a summer or two.) Water lying below the continuous permafrost layer is usually saline or of such poor quality that the costs of drilling, conducting the water through the permafrost, and treatment usually do not justify its use.

In some instances shallow wells placed in the *thaw bulbs* of rivers and lakes can provide a year-round water supply, but the place of usage may be too distant to allow its transport by pipes. A common practice then is to haul the water on sleds or trucks, sometimes as ice but usually in liquid form. An example of how labor-intensive and expensive this method of transport can be comes from the DEW Line radar sites built some years ago across Alaska and northern Canada. An engineer associated with their construction in Alaska claimed that it would have been cheaper to haul water from Fairbanks in chartered aircraft than to pay for the crews hired expressly to sled water in from nearby lakes.[16] Furthermore, the quality of such water changes with the season, deteriorating toward late winter as thickening surface ice rejects salts and organic matter down into the remaining water. By springtime the water in shallow lakes may become undrinkable unless treated.

Water in areas of discontinuous permafrost

Where permafrost is discontinuous, drilling into nonfrozen soil may yield water, and drilling through a permafrost layer may tap ground water below. Supplies may be limited by low rates of recharge—primarily into the nonfrozen areas and mainly in springtime through snowmelt rather than by rainfall in summer—and the water quality may be poor. Hardness is one problem and another is iron, some in part released by thawing of

---

16. Rice and Alter (1974).

peat-rich perennially frozen soil. As northern engineering pundits Eb Rice and Amos Alter stated, "Iron-bearing water plugs pipes and boilers with scale, causes ugly brown stains on plumbing fixtures, turns white laundry amber, turns whiskey violet, and turns housewives livid."[17] One method of avoiding these problems, much used by residents in the outlying areas of Fairbanks, Alaska, is to make do with water hauled in trucks or brought into the home in smaller containers. When a 1960 census taker asked my rural neighbor near Fairbanks if he had running water, the man responded that his water walked in but ran out. Like many of us at that time, he carried water into the house in 5-gallon cans but it left through a pipe leading out from the kitchen sink. People with such rudimentary water supply systems quickly develop excellent conservation habits.

Sewage

The most difficult problem associated with sewage in areas where ground temperatures are below freezing during all or most of the year is transport of the material to disposal or treatment sites more distant than the nearest outhouse. Even those northern communities that have sewer systems usually contain some homes without sewer connections where disposal involves removal in plastic bags or "honey buckets" to locations where final disposition might involve at least primary treatment. If not, the operational philosophy, especially in communities bordering on oceans or rivers, is one of dilution-is-the-solution-to-pollution.

More desirable is primary treatment—the settling or filtering out of solids in tanks or lagoons. Alberta's Department of Environment has found that satisfactorily high-quality effluent drains away from sewage retained in a two-cell lagoon system for at least 24 months.[18] This is a passive approach compared to others that may involve mechanical filtering and possibly chemical precipitation and which permit handling of large quantities of sewage on a time scale of hours to days. Temperature is a factor here since the rate of settling out decreases with lowering temperature, owing to cold water's relatively high viscosity. Governmental rulings may necessitate secondary treatment, that involving chemical or biological processing of sewage prior to its dispersal into the environment. Alaska, for example, has required such treatment at industrial sites such as the camps along the trans-Alaska pipeline, and experience has shown that

---

17. Rice and Alter (1974).
18. Grange and Shaw (1971).

operators of the systems in permafrost areas have had difficulty meeting legal requirements on the effluents.[19]

In northern communities with sewer and water utility systems the common practice is to combine the distribution and collection lines in insulated enclosures called utilidors. These may be large underground tunnels that people can walk through or be much smaller box-like conduits containing sewer and water lines embedded within rigid insulation. These smaller utilidors are sometimes placed above the ground to avoid thawing the frozen soil below, and it is common practice to employ a double-pipe circulating water supply network to keep the temperature in the utilidor above freezing. The cost of utilidors is high, amounting to hundreds or even thousands of dollar per foot. When such a system was installed some years ago in Point Barrow, Alaska, the cost of the multimillion dollar utilidors reportedly far exceeded the value of the homes served.

In populated northern locations like Fairbanks where permafrost is discontinuous and where the population is dispersed over a wide area, most sewage is disposed of in individual septic systems. Conventional systems composed of a leak-proof septic tank leading out to a drain field perform quite well in nonfrozen areas, some operating for decades without attention. Much used in past years are open-bottomed septic pits that sometimes operated for several years before collapsing or needing to be pumped out. These are usually timber cribs but some homeowners have used whatever was at hand, including arrays of punctured 55-gallon drums or old VW buses. In an area such as this where governmental regulation is not stringent, homeowners are able to take innovative approaches to their sewage disposal problems. Incinerating and composting toilets are not uncommon in this area, and one bed and breakfast facility even has a small outlying building where earthworms compost the sewage.

■ Roads and Railroads in Cold Climates

Nowhere are the engineering consequences of thawing permafrost and the annual freeze-thaw cycle in the active layer more obvious than along northern highways. Disturbing to many residents is the sight of an expensive new roadway undergoing failure within a year or two of its completion, or more slowly developing a roller-coaster surface that eventually becomes unusable to traffic. **Figures 5.22** to **5.24** contain examples.

---

19. Zemansky (1975).

Figure 5.22A–B Rollercoaster roadbeds: A) A now-abandoned section of the Alaska Highway 60 miles southeast of Fairbanks, Alaska (photographed 1996), and B) the Copper River and Northwestern Railway 75 miles northeast of Valdez, Alaska, photographed in 1960 by L. A. Yehle, 22 years after maintenance was discontinued (Ferrians et al., 1969).

Figure 5.22C–D C) A warning sign on the Hart Highway between Prince George and Dawson Creek, B. C, and D) a bicycle trail in Goldstream Valley north of Fairbanks so irregular (and wet in places) that it provides cyclists with thrilling rides. Photographed in 1996.

Figure 5.23A–B Longitudinal cracking in roadways: A) a gaping hole in a previously repaired section of the Alaska Highway near Northway Junction, Alaska, and B) Failure of new roadbed still under construction adjacent to a thermokarst lake near Mile 1131 (Koidern River, no. 2 crossing).

Figure 5.23C–D C) Snow-filled longitudinal cracks in the Alaska Highway near Northway, and D) Longitudinal cracking in an abandoned section of the Alaska Highway 80 miles southeast of Fairbanks; photographed in 1997.

Figure 5.24 Upper photographs: Dangerous failures on the sides of Farmers Loop Road that developed within years after its reconstruction in 1994. Lower photograph: A white-painted test section of roadway on the Alaska Highway 80 miles southeast of Fairbanks, photographed in 1997. The smoothness of this test section contrasts with the roughness of the adjacent roadway, attesting to the effectiveness of this method of reducing heat input to a roadway over permafrost. Despite the success, Alaska has not chosen to paint its roads white.

The primary problem is a combination of frost heaving and the thawing of soil below the roadbed that may have been frozen for thousands of years. Inadequate knowledge of cold soil behavior limits society's ability to deal with this problem, but in Alaska at least it appears that a contributing factor is federal and state government planning, financing, and contracting procedures that do not foster high-quality roadway construction.

When constructing a new road over ice-rich permafrost, engineers may choose to aim at avoiding any change in the thermal regime of the subgrade. This preventive approach is practical where mean annual ground temperatures are below −3°C, but is not economically feasible at warmer locations. The approach involves placing very thick earthern fill in the roadway embankment and perhaps putting expensive rigid insulation below or in the embankment.[20]

Another widely used approach is to assume that permafrost will thaw and then attempt to design the road embankment to lessen the damage when subsidence develops. Following either the preventive or mitigative approaches, engineers generally have found it wise to preserve existing vegetative cover. In extreme situations they may elect to minimize the formation of segregation ice by replacing fine-grained soil in the subgrade with rock or gravel fill before placing the road embankment.

Even if not subject to subsidence by the thawing of permafrost, roadways constructed in areas of annual freeze and thaw undergo damage through cryogenic action. This is a widespread problem because roughly half of the earth's land surface experiences the temperature regime that gives it an active layer. Owing to the comparatively high thermal conductivity of the material composing a roadway, and the fact that roads are generally kept snow-free in winter, the active layer within a roadway is typically thicker than that in undisturbed locations to either side, and thermal gradients are likely to be steeper. Consequently, cryogenic suction draws water up through the subgrade and perhaps into the road embankment in winter, most effectively where the soils are fine grained. The results can be substantial frost heave during freezing and, later on, reduction in the road embankment's or subgrade's bearing strength when springtime thawing occurs.

A dramatic example of frost heaving comes from a roadway constructed on thick loess in southern Fairbanks, Alaska. In 1983 engineers dug up 2.5-meter deep sections of the subgrade in order to put in culverts and utility lines, and then replaced the excavated fine-grained material with gravel. Viewing the depressions that developed at the culvert and utility locations during the following winter, the engineers thought that severe settling had occurred and perhaps the culverts had collapsed. Later they discovered that no settling or collapse had taken place; the apparent depressions at the culvert locations instead were caused by greater heaving upward of the entire road elsewhere, by nearly 40 cm (15 in).[21]

20. Johnson et al. (1984).
21. Stella (1986).

This example points out that a roadway, be it a railbed or one designed for automotive traffic, is not just a static elongated strip of earthern materials stretched across the landscape. Roadways that freeze and thaw annually, if constructed in part of fine-grained material, swell up as they cryogenically breathe in water and store it as segregation ice, then shrink as they expel the melted ice in springtime, perhaps suffering permanent deformation in the process. Since it is the formation of segregation ice that creates frost heave, those parts of roadways that heave the most are the ones that have the silt-sized soils, copious water supplies and temperature gradients that best promote segregation ice's formation. Recall from Chapter 3 that the thickest layers of segregation ice develop when the temperature regime permits the freezing front to stall. For this reason, roadways located in areas of moderate cold may heave more than those in climates having hard, rapid freezes.

Perhaps the most destructive characteristic of frost heave is differential heave—the changes in amount of heave with position.[22] Demarkations between cut and fill sections of roadways are particularly susceptible, as are any other locations where abrupt changes in soil characteristics or water supply occur. Attentive drivers of bitumen-surfaced roads like the Alaska Highway soon learn to recognize from the topography where the next bump in the road will be. To minimize this problem, road engineers sometimes mix the differing soil types together so as to create gradual transitions. A common practice also is to remove frost-susceptible soils down to a depth approximating 50% of the expected depth of frost penetration. Replacement of that layer with coarser materials minimizes frost heave in the upper part of the roadbed and helps provide uniform support to the top surface of the roadway. Fill materials used for this purpose include sand, and sometimes also peat or bark because these materials do not promote the formation of segregation ice, although they may compress unduly or rot away.

Measurements of embankment deformation conducted on the Qinghai Xizand Highway in China[23] and along the Baikal-Amur Railway in Siberia[24] show that embankments in cold areas tend to settle most in their centers and tend to spread away from the centers. Compatible with these measurements are observations that a common form of deformation of surfaced roads is the development of wide cracks and slumps along the center of the roadway, and also along the edges of the paving.[25] The causes

22. Crory et al. (1984).
23. Jianheng et al. (1993).
24. Dydyshko et al. (1993).
25. Crory et al. (1984).

may be several, but all are related to frost heave in the active layer or thawing of permafrost below.

Differential heave in the transverse direction may contribute to the central sinking and spreading apart of road embankments, as may the tendency for frost heave within a crowned roadway to have an outward component directed away from the road center. Thus when the contained segregation ice thaws in springtime, the material in the road tends to come to rest farther from the center of the road than it lay before freezing.

Frost heaving may also create permanent volume change within the roadbed material and time-dependent changes in the road's bearing strength. During heaving, the formation of segregation ice layers tends to desiccate the nearby soil, thus tending to shrink it. Then if melting is slow enough during the spring thaw, the weight of overlying soil can squeeze water out of the roadway at the rate generated, leaving the soil more compacted than before, but perhaps more permeable because of weaknesses and cracks created at the sites where segregation ice layers had formed. If the soil is above the water table the force of gravity applies a suction that helps drain the water away and also creates greater adhesion between the moistened soil particles. Thus by dewatering it, repeated freezing and thawing can actually improve a roadbed that lies above the water table, but it may undergo settling in the process.[26]

A quite different, adverse modification can result when melting occurs within fine-grained subgrade or road embankment material at a rate producing water faster than it can drain away. The excess water then saturates the soil, filling all pore spaces, and thereby reduces the adhesion between soil particles by fully coating them with water. This pore water shares with the soil skeleton the task of supporting the overlying material, taking on a greater share the more rapid the buildup of the water. The pressure exerted (called the pore pressure) increases until the pore water takes up the full load of the overburden, leaving the soil particles with virtually no adhesion to each other and thus with little bearing strength. As the pore pressure continues to increase, the water may begin to flow upward, and if it does it exerts a drag on the soil particles that tends to lift them as well. When that drag exceeds the pull of gravity the soil loses all bearing strength and becomes "quick," hence the name quicksand.

As the soil in a roadway approaches the quick (zero strength) condition, it is easily damaged by passage of any traffic that flexes the road surface and increases the pressure in the subsurface enough to cause flow,

26. Johnson et al. (1984).

either lateral or upward. If the flow is upward, it tends to lift the fines up through the roadbed and may even jet them out onto the surface in the messy flows northerners refer to as mudboils. A commonly used mitigation procedure is to place a layer of tough cloth-like material called geotextile, *geofabric,* or geosynthetic fabric below the road embankment to strengthen the embankment and prevent the pumping of fines up through the soil; see **Figure 5.25.**[27]

Yet another source of roadway failure is melting of permafrost below the embankment. Bitumen-surfaced roadways absorb more heat than the surrounding terrain, and embankments in general are more thermally conducting, so roadway and railroad embankments in permafrost areas develop thaw bulbs that are prone to collect water. The collected water may adversely affect the bearing strength of the subgrade, and it provides a ready source of moisture to migrate upward during the next freezing cycle.

One way to minimize embankment deformation is placement of very wide, slightly sloping shoulders on roadways. Then any heaving motion is always essentially vertical so that the embankment does not spread. Perhaps more important in permafrost areas is that the wide shoulders help water run off the embankment far enough away from the center that it is less likely to accumulate in the embankment's thaw bulb.

Several Canadian provinces and most northern states, including Alaska, place springtime restrictions on the weight of trucks allowed on surfaced roads because heavy loads severely flex the road surface when the pore pressure in the roadbed is high. The economic consequences are severe, both in the seasonal limiting of truck transport and in the cost of road repair. In 1980 it was estimated that, even with stringent load restrictions, road repair costs in Alaska amounted to 18 cents per mile for the passage of a fully loaded truck grossing 40,000 kg, and that 16,000 automobiles could travel the mile at the same cost. Measurements indicated that the maximum damage was occurring when spring thaw depths were near 1 meter.[28]

▪ Pipelines in Permafrost Areas

In general, builders of pipelines to carry oil, gas, or refined petroleum products hope to build—at the lowest possible cost—lines that will operate effectively for long periods and cause minimum environmental damage

---

27. Balcourt (1982); McFadden and Bennett (1991) pp 189–203, 347–49.
28. Conner, 1980.

Figure 5.25 The use of geofabric to prevent fine-grained soil from rising up into road-beds made of coarser material. Top: geofabric being installed on a four-lane roadway west of Fairbanks in 1996. Bottom: geofabric installed in a private driveway onto Farmers Loop Road north of Fairbanks.

through erosion and spills. The goals of long-term cost effectiveness and avoiding environmental damage closely intertwine because any pipeline design that avoids damage is likely to lead to low maintenance, and that is conducive to cost effectiveness. The environmental impacts of pipelines in permafrost areas arise from two main causes: 1) disruption of the soil's thermal regime by the construction of a pipeline and its associated work pads and roadways, and 2) modifications to the thermal regime by the operation of the pipeline. Approaches to minimizing the first category of disruptions are the same as that for building roads in permafrost areas: seek to eliminate thawing of frozen ground entirely or design embankments in ways that mitigate the damaging effects of thawing when it occurs. The extent to which the operation of a pipeline affects the thermal regime depends on the temperature of the material carried: hot-oil pipelines can cause extensive thawing and settling if buried, lines that carry oil or fuel at low above-zero temperature may have modest effect, and those that carry gas or liquid at subzero temperature can freeze the soil and cause damaging frost heave.[29] The economic costs of pipeline failure can be very high, especially in countries or areas that depend heavily on the export of gas or oil for currency needed to pay for importing food or other necessities.

Table 5.1 lists some of the characteristics of four major pipelines in permafrost areas, in order of increasing temperature of the oil or gas carried. The cold Mastakh-Yakutsk gas pipeline in Siberia is one of the first constructed in cold permafrost (mean annual temperature at Yakutsk is −10.3°C, and the permafrost is 200–300 m thick). Experience with this pipeline indicates the best performance is with buried portions since seasonal temperature variations generate stresses that exceed permissible limits in the aboveground portions of the pipe. The buried portions are within the active layer so seasonal frost heaving and thaw create vertical motions up to 23 cm. Counter to expectations, permanent vertical displacements have not occurred where soils are most susceptible to frost heaving, but rather in desert areas where sand flowing under the heaved pipe prevents it from settling to its former position. The buried cold pipeline itself apparently has little effect on the thermal regime of the nearby soil, but disturbance to the natural ground cover by placement of the pipeline and subsequent traffic has increased the thickness of the active layer by up to 80% along the right-of-way. Ravine erosion, development of thermokarst topography, and slumping of thawed soil in the disturbed area evidently have caused displacements of the pipe and increased the difficulty of travel along the pipeline.[30]

Another gas pipeline complex carries warm gas (20°C at input) across discontinuous permafrost in the Nadym area of Russia. Individual rights-of-way range up to 500 m wide and carry as many as 10 pipes buried 1 m.

---

29. Williams (1989) pp 62–68.
30. Burgess et al. (1993).

Table 5.1 Four pipelines operating in permafrost areas, listed in order of increasing temperature of the oil or gas carried.

Pipeline (Type)	Length km (mi)	Diameter cm (in)	Environment	Burial Status	Operating Temp.
Mastakh-Yakutsk, Russia (Gas)[a] Built pre-1970	380 km (237)	53 cm (21)	Cold Permafrost	Above ground and buried 0.4–0.5 m	−20°C up to +7.6°C
Norman Wells-Zama, NWT (Oil) Built Winters 1983-85 Operating 1985	869 (543)	33 (13)	Discontinuous Permafrost	Buried 1 meter	−2°C input to +7°C along line
Nadym, Russia (Gas) Built winters starting 1971 Operating 1977[b]	Multiple lines totaling 1,600 (1,000)	142 (56)	Discontinuous Permafrost	Buried 1 meter	+20°C input, no refriger-ation along line
Trans-Alaska (Oil) Built 1975-77 Operating 1977[c]	1,287 (800)	122 (48)	Cold Permafrost to Discontinuous Permafrost	½ elevated, ½ buried 3 to 5 m (6 km of buried line refrigerated)	near 60°C input and along line

a. Kamensky et al. (1993).
b. Burgess et al. (1993).
c. Coates (1993).

During the wintertime-only construction, which began in 1971, the builders sought to route the line over the soils least conducive to frost heaving and permafrost melting, but otherwise undertook no mitigative measures—such as attempting to insulate the pipe to protect permanently frozen ground. Postconstruction monitoring of the Nadym complex showed that the clearing of the right-of-way and other human activities related to building and operating the line caused substantial change. In certain areas the clearing caused the melting of ice wedges and the consequent development of thermokarst topography, while in others the clearing caused subsurface permafrost layers to coalesce into a single unit extending up to the bottom of the active layer. In nonpermafrost layers, the active layer increased in thickness far more in dry sandy areas than in boggy ones. Over the course of 20 years, permafrost in boggy areas thawed to depths near 5 meters, whereas the depth of thaw in undisturbed areas increased only slightly (from approximately 1 m to near 1.5 m at one site), perhaps due to climatic warming. Long-term settling due to melting of segregation ice within sandy soils in the Nadym area amounted to only a few tens of centimeters but in ice-rich peaty soils the settling was up to 3 m. During the first few years, thaw processes and the modifications to runoff waters caused the width of the disturbed areas to grow by up to 30%, then after 20 years the width increased another 35 to 60%.

Partly because of their similarities, the Nadym complex and the Norman Wells-Zama oil pipeline, constructed in the Mackenzie River area, NWT, during the winters of 1983–85, have received joint study by cooperating Canadian and Russian scientists. Both lines are in discontinuous permafrost areas, both are buried approximately 1 m and their temperatures are similar, the Norman Wells oil line being somewhat cooler. In addition to attempting routing through the best soils, as with the Nadym pipeline, the construction of the Norman Wells line involved placing beds of wood-chip insulation around the buried pipe to minimize heat flow into the surrounding soil.

Chilling of oil to near –2°C as it enters the pipeline at Norman Wells brings the oil to near ambient ground temperature (-1 to 0°C) and it remains there for the first 20 km of the line. Thereafter, the oil warms to reach 7°C at the southern terminus. Although intended to be thermally passive, the pipeline has caused thawing of permafrost, as has the disturbance to the surface during construction. Heating within the wood chip insulation as it decays also has been greater than expected and has required remedial action. Settling of fill placed in the pipeline trench has exceeded 50 cm in some places and that has required refilling around approximately 100 km of the pipeline. During the early years of operation, other remedial action to minimize erosion has been necessary.[31]

Most famous of all northern pipelines is the trans-Alaska oil pipeline. It has been described as the most ambitious construction scheme in American history, and also as the world's largest private industrial undertaking.[32] In 1969, ignoring all prior experience with constructing and operating pipelines in cold climates, the builders of the hot-oil trans-Alaska pipeline originally proposed to bury the entire 1,287-km (800-mi) length extending across Alaska from Prudhoe Bay to Valdez through a region of continuous

31. Burgess et al. (1993).
32. Coates (1993).

and discontinuous permafrost.[33] Strong stands by federal officials, primarily those in the U.S. Department of Interior, forced the consortium of builders to alter their plan so that 676 km (420 mi) would be built above ground. The basis for enforcing the change was a series of calculations by USGS geologist Arthur H. Lachenbruch showing that heat flowing from the hot (near 60°C) pipeline would create rapid thawing of nearby frozen soil. Lachenbruch calculated that within a few years the thawed region would extend out beyond 10 m in cold permafrost regions and to twice that in the warmer discontinuous permafrost zone.[34] The result would be major slumping in ice-rich areas that would damage the pipeline and create widespread ecological damage along the right-of-way.[35]

Construction commenced in 1975, and the first oil flowed in 1977. Passive refrigerating thermal pilings supported the above-ground pipeline in the warmer permafrost areas but it was possible to use nonrefrigerated pilings in the northernmost cold permafrost area. Refrigeration also protected a total of 6 km (4 mi) of the buried line in places where elevation was not possible for various reasons, for example, to create game crossings. It appears that the elevated portions of the pipeline have held up well, but workers report that substantial lengths of the buried 380 miles of pipe have required repair owing to melting of permafrost and corrosion. Corrosion is known to be the cause of several failures in a nearby pipeline, the buried 8-inch ALCANGO pipeline completed in 1955 to carry petroleum products from Haines, Alaska, to Fairbanks, a distance of 1,000 km (626 mi).[36]

The trans-Alaska pipeline is particularly susceptible to corrosion because it spans the northern auroral zone. Intense (million-ampere) electric currents appear in the auroral zone ionosphere during times of high auroral activity, and these induce strong voltages in the earth's surface below and in any long metal objects such as power lines and pipelines. The

---

33. This was not the first pipeline project on the far northwest portion of the continent; commencing in 1942 the U.S. War Department began the Canadian Oil (Canol) project to bring oil from Norman Wells to Whitehorse. Laid aboveground, the 4-inch Canol #1 line carried oil to Whitehorse for refining into fuel that then was carried by the 2- to 4-inch Canol #2, #3, and #4 lines to Haines, Watson Lake, and Fairbanks, respectively. Canol # 1 was sold for scrap in 1947–48, and Canol #4 was replaced by a buried 8-inch fuel pipeline between Haines and Fairbanks in 1955. Because of corrosion and other problems, these pipelines suffered various poorly documented failures over the years, but the resulting spills were comparatively small and mostly went unheeded. [Coates (1993) pp 71–75, p 213; Mighetto and Holmstad (1997)].
34. Lachenbruch (1970).
35. Coates (1993).
36. Mighetto and Holmstad (1997) pp 210–11.

induction is greatest (up to 1 volt/km) in the general north-south direction—the general orientation of the trans-Alaska pipeline—so the pipeline contains strong electrical currents during times of intense auroral activity.

In an attempt to prevent corrosion, the pipeline builders coated the buried parts of the pipe with electrically insulating material to minimize current directly between the pipe and the surrounding soil, and installed sacrificial zinc anodes to carry the current instead. However, measurements made during the few years following completion of the pipeline indicated that the applied insulating material contained unavoidable imperfections (called holidays) that did allow current between the ground and the pipe over about 0.02% of the pipe's area.[37] Thus the pipe, its zinc anodes, and the surrounding soil became a battery. In itself, this would not corrode the pipe because the holiday areas should act as cathodes, and the only corrosion should be to the sacrificial anodes, which from time to time could be replaced if necessary. However, when the aurorally driven currents in the pipe are in a direction to reverse the role of anode and cathode, the pipe should corrode. Those having seen dug-up parts of the pipe say that they have observed corroded spots at the locations of the holidays.[38] The temperature of the soil surrounding the pipe and the sacrificial anodes is important because if the soil is frozen its electrical resistivity is comparatively high and it becomes a very poor electrolyte. For this reason, it is desirable that sacrificial anodes on pipelines be placed, if possible, in unfrozen ground. See also discussion on next page.

In an early phase of the overall trans-Alaska pipeline project the state of Alaska bulldozed the tundra aside (rather than building over it) to construct a winter haul road along the northern half of the pipeline route. Built at the direction of then-Governor Walter J. Hickel, the roadway was popularly known as the Hickel Highway. Because of its somewhat trenchlike character after thawing of permafrost, some people also called it the Hickel Canal. The road was used only one month and was a financial failure; however, that mattered little because its real purpose was psychological—to destroy the idea that this was pristine wilderness basically untouched by man. After nearly three decades, little sign of the Hickel Highway remains, but it did its job, and public response to it probably did much to promote the effective road and pipeline construction and maintenance work conducted ever since along the route. The interaction between oil industry builders and operators of the system and the federal and state agencies overseeing it continues to be

---

37. Wescott et al. (1979).
38. Sackinger (1980).

Corrosion

Corrosion is the process wherein a substance such as iron (steel) is destroyed at its surface by chemical action that causes the iron to combine with oxygen (or other oxidizing elements like sulfur and chlorine) to form iron oxides. Iron and oxygen have less free energy when combined than the molecules standing alone, and so the process proceeds given the slightest excuse. For the iron, it is a return to the natural state it had before mankind expended energy to purify it.

Iron heated in dry air will oxidize and, as every user of a cast iron skillet knows, iron (steel) corrodes easily when exposed to wet air, and even more easily if the wet air contains salt. Hence, after washing a skillet it is wise to give it an oily coating to inhibit the rusting. In doing this, the user is preventing the skillet from becoming a battery by removing from its surface the water that acts as an electrolyte. Recall that a battery is a device for converting chemical energy to electrical energy, and that it consists of two electrodes immersed in an electrolyte. The physical difference between the electrodes is that one (the cathode) has a higher electron attracting power than the other (the anode). Even though a rusting skillet's surface may look uniform, it is not, and the rust is appearing near places where imperfections give iron atoms a lesser hold on their electrons than those located elsewhere, and each of these places with less holding power is acting as an anode. Having lost some of their electrons at anodic locations, the remaining iron ions drift off into the electrolyte where they combine with oxygen to form iron oxides, i.e., rust. The lost electrons drift through the metal to more attractive locations that act as cathodes where they react with oxygen and water to produce negatively charged hydroxide ions (OH^-), leaving the cathode area unscathed. So when it rusts (corrodes), iron typically becomes pitted; the pits are where the surface takes on an anodic character. The skillet battery or other corroding object consisting of but one element produces minimal voltage between the anodic and cathodic locations because the differences in electron attracting power are very slight. However, substances differ in their electron attracting power, so a practical battery circuit has an anode of one substance and a cathode of another. A battery with a zinc anode and an iron cathode produces a voltage of about 0.3 volts, and it operates by corroding away the zinc anode. This is the reason for connecting sacrificial zinc rods to steel pipelines; instead of the pipeline corroding by parts of it becoming anodic and other parts cathodic, the pipe acts as the cathode and remains intact, whereas the zinc forms the anode and gets eaten away. If a large enough external voltage is impressed on the zinc-iron battery in the direction to reverse the flow of electrons, the zinc becomes the cathode and the iron becomes the anode, so then it chemically erodes.

confrontational, but anyone traveling along the trans-Alaska pipeline has to be impressed by how little environmental damage is obvious to the eye. Most of that appears to be associated with roadways and support facilities rather than with the pipeline itself.

■ Chapter Summary

Hunter-gatherer societies in cold lands have experienced little trouble with perennially frozen ground, and have sometimes used it to advantage, such as for food storage. When gold miners came into Alaska and northern Canada, they learned by trial and error how to thaw permafrost, and the larger mining operators developed fairly sophisticated techniques. Even so, the cost was high, and dealing with frozen ground accounted for half of the cost of mining in the Fairbanks and Nome areas. During past decades, using techniques developed in warmer areas, government agencies and others have erected many buildings in the north only to see them fail because of the thawing of permafrost beneath the foundations. Some home owners have tended to deny the existence of a problem or attributed structural failure to other causes, but most northerners have developed a proper awareness of the hazards of building on ice-rich permafrost and have learned to deal with it. However, people erecting structures on unfrozen soil that overlies permafrost need to realize that foundation problems can develop even if the permafrost table is well below but still within a few tens of meters of the ground surface. Water supply and sewage disposal are critical problems in the continuous permafrost region and a very serious one even in many areas where the permafrost is discontinuous. Frozen ground limits the availability and quality of water, and it makes sewage removal and treatment both difficult and expensive. Another very serious and expensive consequence of seasonally and perennially frozen ground is its effect on transportation facilities—roads, railways, airport runways, and pipelines. Pipeline builders have achieved reasonable success in maintaining pipelines across discontinuous and continuous permafrost areas, but construction and maintenance costs have been high. Less successful are attempts to build surfaced roads that last for more than a few years, and this is an arena in which much needs to be learned.

Permafrost in Transition

▪ Introduction

Freezing and thawing ground, and even ground that remains perennially frozen is always changing. The ground swells and shrinks as the temperature varies, soil particles alter their shape, and the water contained in the ground moves around. And so when the climate changes, the nature and extent of frozen ground changes as well.

▪ Climate Change, Evidence of and Causes

Long-term trends and sudden events, either natural or man-made, can alter ground temperatures. Forest fires and land-clearing operations, or even earthquakes and ash injected into the high atmosphere by erupting volcanoes sometimes create changes in ground temperatures by altering the local ground cover or disturbing preexisting drainage patterns. Comets or meteorites exploding in the atmosphere or their impacts on the ground can have local or perhaps even global effects.[1] Most profound, however, are the changes in climate that affect broad regions or the entire earth, and these are due largely to variations in the amount of solar energy reaching the ground. The main controls on climate are 1) the amount and direction of solar energy arriving at the top of the atmosphere, 2) absorption of incoming radiation

---

1. An example is the mysterious explosion of June 30, 1908, over the Stony Tunguska River in central Siberia that felled and burned forests within an area 45 km in radius. It probably was an ice-rich comet exploding in the atmosphere. About the same size as the Mt. St. Helens eruption of 1980, it released as much energy as a ten-megaton atomic blast.

by the atmosphere and the ground, 3) reflection of incoming solar radiation from clouds, snow, soil and water surfaces (the earth's albedo), 4) the distribution of continents and oceans, and 5) topography: the elevation and degree of ruggedness of the land and its relation to other topographic features.

During long-past eras—tens to hundreds of millions of years ago—the changing distribution of the continents and oceans and the topographic changes did alter the climate, but within the recent few million years these controls have changed little and therefore have not been significant. Changes in the earth's albedo occur only in response to other variables, and so the albedo is not an independent controller of climate. That leaves but two independent causes of climatic change: variation in the solar energy impinging on the atmosphere, and variation in the atmosphere's ability to absorb and reflect that energy. During the past 150 years, humans have enhanced the atmosphere's ability to absorb solar energy by injecting industrial gases that have increased the efficacy of trapping of solar energy by the *greenhouse effect*, discussed in the following.

▪ Variation in the Amount and Direction of Solar Energy Impinging on the Top of the Atmosphere

Solar energy arrives at the earth in two forms: radiated sunlight and a combination of kinetic and magnetic energy carried by the solar wind. The solar wind delivers energy to the earth at a rate only roughly one-millionth that of sunlight, but even so it is possible that solar wind variations have significant effect on weather and climate because of the special nature of the energy carried. It is not easy to sort out solar radiation effects and solar wind effects because it is now known that both vary with solar activity.

Variation in the amount of radiated solar energy (sunlight) arriving at the top of the earth's atmosphere now appears due to two primary causes: variable output of the sun, and changing sun-earth geometry. The sun-earth geometry changes because of variations in the earth's orbit that are described by three parameters:

1. Eccentricity. The earth's orbit gradually changes from circular to elliptical and back to circular with a period of about 100,000 years. During this shift the distance between the earth and sun changes enough to cause a variation of about 0.1% in the intensity of solar light.

2. Obliquity. The earth's spin axis rolls in regular fashion that changes the direction between the earth's equator and the plane of its orbit around the sun (the ecliptic) over the range 22.1° to 24.5° with a period of 42,000 years. The change alters the latitudinal distribution of sunlight.

3. Precession. The gravitational pull of the sun and the moon on the earth's equatorial bulge causes the equinoxes to occur earlier each year and the spin axis to wobble with periods of 19,000 and 23,000 years. Like the change in obliquity, the precession changes the amount of sunlight arriving at a given latitude at a particular time of the year.

Statistical analyses of climate-sensitive fossil deposits found in ocean floor boreholes and analyses of oxygen isotopes in ice cores acquired from deep holes in Greenland and Antarctica reveal that during the past 500,000 years climatic changes have tended to occur at 23,000-year, 42,000-year, and 100,000-year intervals, so it is evident that the regular long-term changes in sun-earth geometry have influenced climate. These orbital variations (sometimes referred to as Milankovitch cycles after the man who, in 1924, first proposed their impact on climate) appear to be important drivers of the oscillations between epochs of glaciation and interglacials during the Pleistocene.[2]

Satellite observations during the past two decades show that the sun's output of radiant energy (electromagnetic radiation of all wavelengths, including visible sunlight—far and away the main energy output) varies by about 0.1% during the 11-year sunspot cycle. Extrapolation of the satellite data by means of sunspot records has allowed reconstruction of the sun's radiant energy output back to the year 1600. Global temperature records, also mostly reconstructed back over the same 400 years, show changes that closely follow the sun's radiant energy output—at least up until about 1860, when man's industrial activities began to alter the picture. Since then, increasing solar output has been only partly responsible for a continuing warming trend; estimates range from 10% upwards to more than half of the change in temperature.[3]

---

2. Hays et al. (1976); Hays (1977).
3. Crowley and K.-Y. Kim (1996); Kerr (1996).

▪ Variation in the Atmosphere's Ability to Absorb and Reflect Sunlight

Approximately 6% of the sunlight impinging on the top of the atmosphere reflects back into space from molecules in the clear atmosphere, another 25% reflects from clouds, and 4% reflects from the earth's surface, making the total part reflected about 35%. Called the *planetary albedo*, this fraction of the incoming light merely bounces off the earth system without modification other than receiving a change in direction. The remaining 65% of the incoming radiation does undergo change—change in wavelength or conversion to kinetic energy that warms the earth and its atmosphere and drives the circulation of the atmosphere and oceans. This portion of the incoming radiation causes the earth system to warm or cool until the system can come to equilibrium by radiating back to space an amount of energy exactly equal to that entering.

In the high atmosphere above 60 km, oxygen absorbs about 0.02% of the incoming radiation, ozone absorbs about 3% in the stratosphere (altitude 20–25 km), and a combination of water vapor, carbon dioxide, minor constituents, and dust and clouds in the lower atmosphere absorb another 15%; thus the atmosphere absorbs 18% of the incoming radiation. The earth's surface absorbs the remaining energy; that is, the 46% left over after subtracting the albedo and the part absorbed by the atmosphere. Both the atmosphere and the earth's surface radiate the absorbed energy, but at wavelengths far longer than that of the incoming solar energy because temperature determines the wavelength, and the cooler a body the longer are the wavelengths of its radiated light. The sun is hot and so emits white light (the light of the visible spectrum), but cooler objects, those at temperatures found near the surface of the earth, emit infrared radiation. Because the atmosphere is partially opaque to the long-wave infrared radiation from the earth and from within the atmosphere itself, the atmosphere as a whole absorbs the infrared energy and heats up. This is crucial, for were it not for this opaqueness of the atmosphere to infrared light, the earth's global temperature would be near −23°C instead of the approximately 15°C temperature we have now (assuming that the albedo remained at 35%). This is the "greenhouse effect," so called because that is the way a greenhouse operates: shortwave (white) light penetrates the glass and becomes absorbed by plants and soil within; these then radiate infrared-wavelength energy that cannot pass through the glass, and so the greenhouse heats up.

One of the main radiation absorbers in the atmosphere is ozone (O_3), a highly reactive molecule formed in the stratosphere by ultraviolet sunlight action on normal oxygen (O_2). Ozone does double duty by absorbing strongly at the two extreme ends of the solar spectrum. It shields the lower atmosphere and the earth from the highly damaging incoming solar ultraviolet radiation, and it also contributes to the greenhouse effect by absorbing infrared radiation coming up from below. This absorption is in an important wavelength region where radiation from the earth's surface peaks, but where other strong absorbers (water vapor and carbon dioxide) and weak absorbers (methane, nitrous oxide, carbon monoxide, and nitric oxide) absorb poorly. Thus there are two reasons for concern over variations in the atmosphere's ozone content: 1) its role in the shielding from damaging ultraviolet radiation, and 2) its role in moderating global temperature.

Two other important contributors to the atmospheric greenhouse effect are carbon dioxide and water vapor, both end products of combustion, and both such effective absorbers of infrared radiation that they strongly control global climate. Furthermore, life on the planet's surface links so intricately with atmospheric carbon dioxide and water that any changes to one of the three can easily feed back to alter the other two in ways that might create irreversible local or global changes in climate. When photosynthesizing, plants extract carbon dioxide from the air at rates that may be limited by the concentration of carbon dioxide in the air, and that also depend on sunlight intensity, temperature, and humidity. In addition to oxygen, plants also give off carbon dioxide, but on the whole at a lesser rate than they assimilate it, so living plants constitute a sink for carbon dioxide. Animals produce it, and microorganisms produce a great deal more by consuming organic matter. Burning processes—the burning of vegetation or of fossil fuels like coal or petroleum—return carbon and oxygen to the air in the form of carbon dioxide, and thus act to increase its concentration. If the burning is of fossil fuels, the atmosphere receives back an injection of carbon, hydrogen, and oxygen that was removed thousands to millions of years ago.

Other greenhouse gases of lesser importance are methane, nitric and nitrous oxide, carbon monoxide, and various compounds of chlorine, fluorine, and carbon (chloroflourocarbons, or CFCs). They exist naturally and human industrial activities produce them. As well as absorbing infrared energy, the chlorofluorocarbons attack ozone and remove it from the stratosphere, the main reason why governments have instituted controls on CFC emissions.

Acting in opposition to the greenhouse gases by blocking out incoming solar radiation are concentrations of dust and other particulate matter that contains compounds such as sulfur dioxide. Some concentrations derive from volcanic eruptions, and some from comet and meteorite debris; since the beginning of the Industrial Revolution in the 1760s an increasing amount comes from industrial activities, especially since petroleum came into the picture in 1859.

▪ The Known History of Climatic Change

Each passing year brings new information on the details of past climatic change, mainly because of improvements in dating methods and increasingly sophisticated techniques for determining past climates and measuring the processes that control climate and weather. For example, only during the past two decades has it been possible, using satellites, to obtain highly accurate measurements of the sun's energy output, and consequently only within the past few years has it become obvious that variability in the sun's output is a significant factor in climate control. As this new information becomes integrated into the overall understanding of the causes of climate change, it improves predictions.

Table 6.1 presents a summary of known climatic events and trends during the past billion years or so. While examining the table it is useful to refer to **Figure 6.1** which provides a quick overall view of how global temperatures have changed during the past 500,000 years, and how the various changes relate to identified glacial and interglacial stages. Because dating techniques are improving and new knowledge is rapidly accumulating, it is likely that some of the information in Table 6.1 and Figure 6.1 will soon become outdated, so perhaps a person should look more at the trends displayed than the exact details presented.

A striking characteristic of the curves in Figure 6.1 is the lack of any plateaus; the curves show that the global temperature is always either increasing or decreasing—it never stays constant. The implication of this behavior is that the interlinked collection of climatic driving forces and the responding transfers of energy within the oceans and the atmosphere have sufficiently different time constants and inertias to prevent ever achieving a steady-state balance. Even the long periods of cooling during glaciations (top panel of Figure 6.1) that evidently are driven mainly by changes in the eccentricity of the earth's orbit suffer temporary interruptions, with the result that the last major glaciation, the Wisconsin, contains four subglaciations separated by intervals of relatively short-term warming. Then when the major glaciations end, the typical pattern is one of rapid warming lasting upwards of 10,000 years. Relatively violent fluctuations in mean annual air temperatures may accompany a switching

Table 6.1 Past climatic events and trends, along with discussion of causes.

Event or Trend	Discussion—Probable Cause
600-800 million years ago—Late Precambrian glacial age. **About 300 million years ago**—Permocarboniferous glacial age. **Next 200 million years**—warm period, probably warmer than present. **55 million years ago**—long-term cooling trend began, continuing to present. **10 million years ago**—glaciers begin appearing in high-latitude mountainous regions, rapid expansion of Antarctic Ice Sheet.	Believed caused by major changes in ocean circulation patterns as the continents drifted. Also suggested as contributing to onset of the Permocarboniferous glacial age is the increase in woody plants that would remove CO_2 from the atmosphere and thereby cause global cooling.[a] Long-term solar variability is another possibility.
5 million years ago—A second expansion of the Antarctic Ice Sheet. **3 million years ago**—Continental ice sheets first appeared in northern hemisphere	Perhaps related to the ongoing changes in continental and oceanic patterns but other variations such as change in solar output might have played a part.
Past 500,000 years—Oxygen isotope ratios and concentrations of small siliceous and calcareous organisms (radiolaria and foraminifera) fossils in cores taken from deep-sea sediments and ice cores show a 100,000-year pulse in climate with slowly changing epochs of increasing glaciation followed by abrupt climatic warmings during the Pleistocene.[b] See top panel of Figure 6.1. Also much shorter variations occurred during the early part of this era, and also during the past 10,000 years, with cycles near 6000, 2600, 1800 and 1400 years.[c]	Spectral analysis of the isotope data showed peaks at 100,000, 43,000, 24,000 and 19,000 years corresponding to the periods of changes in sun-earth geometry, clearly showing this to be the primary cause of the cycling between glacial and interglacial epochs. The shorter cycles may be interplay between longer cycles or perhaps due to solar variability.
Past 200,000 years—Interglacial sea levels measured in Barbados show peaks 193,000, 121,000–126,000, 104,000 and 83,000 years ago corresponding to peaks in northern hemisphere summer solar radiation input calculated from changing sun-earth geometry.[d]	Both orbital eccentricity and precession are involved, and calculations indicate the high importance of summer radiation at low latitude.

table continued on next page

a. Retallack (1977).
b. Imbrie and Imbrie (1979) p 146.
c. Kerr (1998b); Oppo et al. (1998).
d. Imbrie and Imbrie (1979) p 146; Edwards et al. (1997).

Table 6.1 *continued*

Event or Trend	Discussion—Probable Cause
~130,000 years ago—Warming trend begins in interior Alaska.	Prior to warming, the temperatures in the area were –8° to –10°C lower than at present.[e]
125,000 years ago—Interglaciation peaked in interior Alaska, producing the Eva Forest Bed near Fairbanks.[f] Then the last (Wisconsin) glaciation began. It involved four substages. The last peaked about 18,000–20,000 years ago. Within the Wisconsin 24 abrupt climate warmings (called Dansgaard/Oeschger events) have been identified.[g]	At peak of the last substage, sea surface temperatures averaged 2.3°C colder than today,[h] but temperatures over Greenland were as much as 21°C lower, and the global temperature averaged 8°C lower.[i]
70,000 years ago—Measurements of nitrogen and oxygen isotope ratios indicate a *Dansgaard/Oeschger event* in which the surface temperature in Central Greenland rose by 16 ± 2 centigrade degrees.[j]	Dansgaard/Oeschger events are characterized by rapid changes in isotopic composition lasting hundreds to thousands of years.
Past 45,000 years—*Heinrich events* (pulses of ice-rafted detritus into the North Atlantic that indicate rapid melting of adjacent continental ice, mostly from Laurentide Ice Sheet,) show a periodicity of 6000 to 10,000 years (about 8400 years). 21,000 ± 3,000 years ago—Last glacial maximum, with tropical latitude temperatures cooler by 2°C over oceans and 4–5 °C cooler over continents.[l]	Suggested cause is variation of incoming solar energy at low latitude affected by orbital eccentricity but mainly caused by harmonics of interlinked forcing by the 22,000 and 19,000 year precessional variations that alter zonal winds and ocean currents—a sort of extremely long-term El Niño, as it were.[k] Could also be due to or affected by variability in solar output.
14,672—Start of sudden warming called the Bølling Transition. Within several decades the Greenland summit warmed by 9 ± 3°C, and within 200 years sea ice melted to about half its present extent.[m, n]	Identified as a Dansgaard/Oeschger event and corresponds to a Heinrich event. This kind of change also is sometimes referrred to as a deglaciation climate reversal.

e. Péwé et al. (1997).
f. Stocker (1998).
g. Péwé et al. (1997).
h. CLIMAP Project Members (1976).
i. Cuffey et al. (1995).
j. Lang et al. (1999).
k. McIntyre and Molfino (1996).
l. Bard (1999).
m. Hays (1977).
n. Severinghaus and Brook (1999).

Table 6.1 *continued*

Event or Trend	Discussion—Probable Cause
14,672—10,000 years ago (dated on basis of annual layering in ice cores from the Greenland Ice Sheet Project 2 [GISP2]—abrupt climate oscillations observed in Europe, Greenland and central North America involving events identified as the Bølling-Allerød warm period (~14,500 to ~13,000 years ago), an oscillatory event called the Intra-Allerød cold period and also the Gerzensee-Killarney Oscillation, lasting a few hundred years, then the Younger Dryas cold period (~13,000 to ~11,500 years ago) followed by the Preboreal Oscillation lasting a few hundred years.[o]	Suggested to be caused by broad climatic variations acting in north or south hemispheres that are propagated globally by the atmosphere. Correlations between oxygen isotope ratios and air temperature indicate that mean annual air temperatures may have varied as much as 6°C during this oscillatory epoch.
7,000 years ago—long-term cooling begins; see central panel of Figure 6.1	Follows the Postglacial Climatic Optimum when global temperature was 2°C higher and rainfall was greater than today.[p] A Heinrich event occurred at about this time.
10,500, 8000, 5300, 2800, and 250 years ago—glaciers were at maximum expansions. Expansion intervals lasted up to 900 years and subsequent contractions up to 1750 years, indicating a little ice age cycle of about 2500 years. The last such expansion is known as the Little Ice Age. See central panel of Figure 6.1	The Little Ice Age temporally encompasses the Maunder Minimum, 1645–1715, when no sunspots were seen and which correlates with a time of low solar output. Thus, cause of the 2500-year cycle appears to be solar variability.
Past 2000 years—Tree-ring growths indicate irregular cycling between dry and wet periods in North America with periods ranging from 30 to 300 years.	Observations of auroras indicate solar variability with periodicities of about 80 and 250 years, in the same range and therefore the most likely cause.[q]
Past 400 years—Northern hemisphere surface temperatures closely track solar light output, showing variations of about 1°C while the solar output fluctuated up and down but generally increased by 0.3%. See bottom panel of Figure 6.1	Indicates detailed connection between solar variability and climate on time scales of decades to hundreds of years.

table continued on next page

o. Yu and Eicher (1998).
p. Imbrie and Imbrie (1979).
q. Davis (1992) p 61.

Table 6.1 *continued*

Event or Trend	Discussion—Probable Cause
1816—"The year without summer," when summer temperatures plummeted in the northern hemisphere. It was the worst of several years from 1812 to 1817.[r]	Three major volcanoes erupted during this period to load the atmosphere with dust that blocked out solar radiation. A similar cold spell occurred in 1785 following eruptions in Japan and Iceland. Also having worldwide effect were the 1883 Krakatoa and 1912 Katmai eruptions.
Past 120 years—Subsurface temperatures in tropical and subtropical parts of ocean basins vary by 0.1°C in time with 11-year sunspot cycle.[s]	Clear evidence of solar variability influencing climate on a time scale of a few years.
Between 1860 and 1940 global temperature rose by about 0.6°C. Then a cooling trend began and global temperature fell 0.3°C by 1975,[t] but overall there has been a gain of about 0.5°C during past 100 years.	During this period atmospheric levels of carbon dioxide, methane and nitrous oxide concentrations have increased. It is suggested that by 1970 about half of the temperature increase was due to the increasing greenhouse effect, and that after 1970 two-thirds of the rise was due to the increasing greenhouse effect.[u]
Past 100 years—Sea level has risen 15–30 cm[v]	Indicates melting of mainly glacial ice, but is partly due to thermal expansion of ocean water.
Past 50-100 years—Permafrost temperatures in northern Alaska boreholes have warmed by 2 to 4°C. Permafrost is thawing at various Alaska locations.[w]	
Past 30 years—Winter temperatures in northern North America and Asia have increased as much as 1.5°C per decade.[x] General retreat of glaciers in northern Alaska also indicate climatic change during this period.[y]	

r. Hughes (1976).
s. Crowley and Kim (1996); Kerr (1996).
t. Imbrie and Imbrie (1979) p 180.
u. Crowley and Kim (1996); Kerr (1996).
v. Ibid.
w. Ibid.
x. Ibid.
y. Rabus et al. (1995).

Table 6.1 *continued*

Event or Trend	Discussion—Probable Cause
1977—Reported step increase by 2°C in air temperatures in Alaska; also step decrease in sea ice cover of Bering Sea. Increasing precipitation in western Alaska.[z]	
1997—The hottest year of the century so far; the four hottest years on record were in the 1990s; nine of past 11 years were warmer than any on record.[aa] As the end of the century neared, the trend continued.	

z. Zhang and Osterkamp (1993); GCASR (1995).
aa. Kerr (1998a).

from a glacial to an interglacial interval, as happened 14,000 to 10,000 years ago; see Table 6.1. These oscillations appear to be caused by or much affected by oceanic and atmospheric circulation patterns.[4]

The earth is now in one of these rapid warming episodes, the Holocene, but even this has had its intervals of cooling, as the central and lower panels of Figure 6.1 and some of the entries in Table 6.1 demonstrate. Overall, however, global temperature is on the rise, and the change has been especially rapid during the past 100 years. The change is so rapid that in the course of one human lifetime sea level has risen enough (15–30 cm) to be measurable and to contribute to the loss or deterioration of some properties fronting on the sea. Perhaps half of the rapid warming is due to human activities, and the other half now appears to be due primarily to the changing output of the sun.

The big question is: Will the warming trend continue? A forecast based only on the changes in sun-earth geometry and incorporating the 2,500-year cycle (perhaps due to solar variation) calls for gradual warming during the next 1,000 years, then 22,000 years of cooling until the planet is in the depths of a new glaciation.[5] On the other hand, it seems that the near future is likely to be determined mainly by a combination of short-term variations in solar output and man-made impacts. Dominating the system might be the man-made impacts: increases in atmospheric greenhouse gases, increases in atmospheric particulates and ozone-destroying chlorine

4. Cane (1998); Steig et al. (1988); Stocker (1988).
5. Imbrie and Imbrie (1979) p 184.

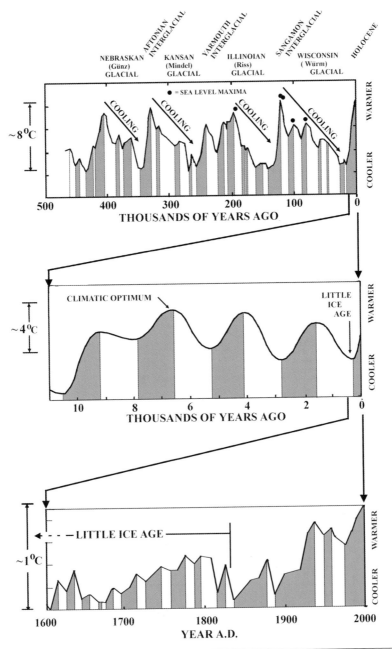

Figure 6.1 Approximate temperature variations during the past 500,000 years deduced from observations of various indirect indicators of temperature. The top panel is actually a plot of oxygen isotope ratios measured by Hays et al.(1976); the second panel is a plot of temperature as estimated from geological records of mountain glaciers and fossil plants [Figure 43 of Imbrie and Imbrie (1979)], and the bottom panel shows reconstructed global temperatures based on data presented by Crowley and Kim (1996) and Kerr (1996).

compounds, removal of tropical forests that serve as sinks for carbon diox-ide,[6] and alteration of the albedo by desertification, agriculture, and urban-ization. Predictions based on the assumption of a doubling of carbon dioxide during the next 100 years suggest an additional rise in global tem-perature of 3°C, along with a rise in high-latitude temperatures by approx-imately 6°C. These predictions do not take into account variation in solar output, and that also has increased during the past 200 years.[7] Even if the man-made impacts do not generate more than minor modifications to natural climate, it is certain that the climate will change, and changes even more dramatic than those of the past 100 years probably will occur.[8] Of major concern is the possibility that the man-made impacts are adding to natural trends in a way that is triggering more rapid and prolonged warm-ing than would normally occur. Perhaps the most serious possible conse-quence in the long run could be enough melting of the global ice reserve to endanger virtually every coastal city in the world.

▪ Response of Permafrost and the Active Layer to Change

Because perennially frozen ground requires soil temperature at or below the melting point of ice, climatic or other changes that alter ground tem-perature generally modify the distribution of permafrost, and are likely to change the thickness of the active layer as well. The rate of heat flow from inside the earth varies considerably with location but only slowly with time. It is minuscule, only about 1/25,000 of the rate of energy arriving from the sun, so it is the solar energy that determines near-surface ground temperatures. Thus, changes in ground temperature start at the top and work their way down, proceeding more slowly with depth. A warming of the ground may melt the top of the preexisting permafrost layer, causing the depth to permafrost to increase. Continued warming causes addi-tional thinning of the layer by melting of the bottom as well, because the warming moves down through the permafrost layer and eventually brings the temperature at the base above that required for the formation of ice. Similarly, cooling of the ground is likely to thicken the permafrost by bringing its top closer to the surface, and if the cooling persists long

---

6. Grace et al. (1995).
7. GCASR (1995).
8. Overpeck (1996).

enough it will cause additional freezing at the bottom of the layer. Where the permafrost is thick, the time scales involved are long. At the temperatures now pertaining there (-11°C), it would take 500,000 years to form the 600-m thick layer of permafrost beneath Point Barrow, Alaska, and if the temperature were a few degrees lower the time required would be about 50,000 years.[9]

Air temperature variations associated with climatic change can also alter the thickness of the active layer. Because it depends on soil water content and on snow and other ground cover as well as on air temperature, the thickness of the active layer may respond in complex fashion to any change that alters ground temperatures or thermal characteristics.[10] In regions of discontinuous permafrost or where the top of the permafrost is well below the bottom of the active layer, long-term increases in air temperature generally will decrease the thickness of the active layer, and air temperature decreases thicken it. Where the top of the permafrost corresponds to the bottom of the active layer, it is more likely that air temperature increases will thicken the active layer, and decreases make it thinner. However, by affecting the nature of vegetation or the amount and timing of snowfall, a changing climate might thicken the active layer in some areas and thin it in others. For this reason, changes in near-surface mean annual ground temperature do not always track changes in mean annual air temperature, with the result that even if the mean annual air temperature is rising, some locations might show decreasing mean annual ground temperature.[11]

Concurrent with these changes to the distribution of permafrost and the thickness of the active layer are variations in temperature within the frozen ground that dramatically alter the thermal and mechanical properties of the material—even without melting. Recall that frozen soil within a few degrees of the melting point, 0°C, contains liquid water in amounts that depend on both the temperature and the size of the pores in the soil. Because the thermal and mechanical properties of liquid and frozen water

---

9. Osterkamp and Gosink (1991); Lunardini (1993).
10. One example of the complexity involves the movement of soil moisture into snow cover and to the air above. Measurements in the Fairbanks, Alaska, area have shown that the uppermost soil layer dries during the winter months, going from 30% moisture content dry weight to about 10%, and that the accumulation in the snow amounts to about 10% of water in the snow pack [Trabant and Benson (1972)]. This transfer affects the thermal characteristics of both the soil and the snow cover [Friedman et al. (1991); Sturm et al. (1995)], and these in turn affect the thickness of the active layer.
11. Zhang and Osterkamp (1993).

differ greatly, it is little wonder that radical changes in these properties occur with variation in temperature, particularly in the range –2° to 0°C. The thermal conductivity of ice is four times that of water, and the heat capacity of ice is but half as much, so as frozen soil nears the melting point its ability to retain energy rapidly rises, and its ability to transmit energy declines. The hydraulic conductivity increases as frozen ground nears the melting point (see Figure 3.15), so frozen ground's ability to transmit water increases as the temperature rises.

Of particular interest are the mechanical properties of frozen soil when the ground temperature rises because the strength of the soil declines, making it less able to support loads, and perhaps become unstable if on a slope. Frozen soil is stronger than unfrozen soil, and also stronger than ice. The frozen soil's strength comes partly from the cohesion of the ice matrix and partly from frictional resistance between the soil particles, but it appears that these sources of strength are somewhat interdependent and not necessarily additive in simple fashion. An example is frozen sand, which has a compressive strength (ability to bear load) greater than the sum of the strengths of the ice and the soil matrix.[12] Probably affecting the strength and other properties of frozen soil are the films of adsorbed water that form interfaces between ice and the soil particles, between ice grains, and at boundaries between ice inclusions and the frozen soil matrix, so the situation is complex. Furthermore, it can change with time because unfrozen water within the soil can move from one place to another, and the amount that is present at each location critically affects the strength there.

When burdened with a sudden load, frozen soil has high strength and therefore resists deformation until it fractures. Like pure ice, frozen soil is brittle, but also like ice, it easily creeps when subjected to slow loading and will flow even if the load is very small. The motion takes place within the ice matrix and it involves melting of ice at locations of highest stress, flow of the resulting water to locations of lower stress (where the free energy is lower) and then refreezing. This is the process of pressure-induced regelation discussed in Chapter 3. In pure ice, slippage occurs most easily along the wrinkled sheets of water molecules oriented perpendicular to the C-axis (the axis pointing along the direction of heat flow when segregation ice forms). In this regard, even a single ice crystal behaves like a pack of cards, undergoing deformation most easily along the planes between the

---

12. The discussion here of the mechanical properties of frozen soil closely follows that of Williams and Smith (1989).

individual cards. This process, called basal glide, takes place about 10 times more easily than creep in the direction along the C-axis, i.e., perpendicular to the card deck.[13] Crystals in a frozen soil probably tend initially to be oriented randomly, but as creep takes place, the crystals reorient to point more uniformly so that the frozen soil undergoes creep more easily as time goes by.

Soil grain size is important because it helps determine the amount of water present in the pores at a given temperature, and the grain size also affects the frictional resistance to deformation. However, temperature is the crucial parameter. As the temperature of frozen soil approaches 0°C a higher proportion of water occupies the pore spaces, which weakens the soil so that the rate of creep increases. The change in strength is most dramatic in the temperature range −2° to 0°C; well below −10°C, temperature variations have comparatively little effect.

The strength of frozen soil increases with increasing particle size because the coarser the soil the less adsorbed water it contains. Another factor influencing the strength of frozen soil is the amount of ice it contains within the pore spaces. The strength of a soil increases with increasing ice content until enough ice is present to fill the pore spaces completely. Higher ice content then reduces the strength because the ice framework takes up more of the load initially borne by the soil matrix, and the ice is more subject to creep than the soil matrix. Soils with laminar structures, such as layers of segregation ice, are weaker than soils without. When subjected to shear, frozen ground tends to fail first along any ice laminations.

Because the mechanical properties of frozen ground radically change as the temperature approaches the melting point of ice, consequences of a warming trend are likely to include increased rates of solifluction on slopes and sudden collapse of slopes that may have been stable for thousands of years. Roadways and foundations may settle as the frozen ground beneath them weakens, adding to the subsidence that occurs due to the melting of segregation ice contained in the soil.

If the current warming trend continues, then the extent of permafrost will decline. The global temperature increase expected to accompany a doubling of atmospheric carbon dioxide over the course of 100 years is predicted to cause the dramatic poleward retreat of permafrost depicted in **Figure 6.2**, which involves elimination of virtually all permafrost within the present discontinuous zone. The permafrost in this zone is warm and,

---

13. Hobbs (1974) pp 274–321.

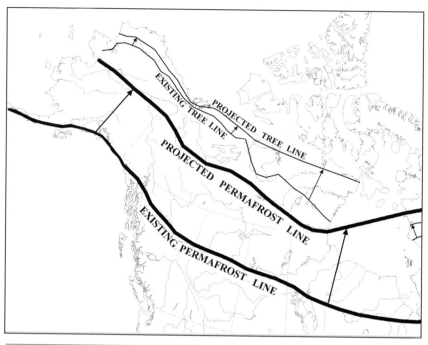

Figure 6.2 Predicted northward movement of the permafrost boundary and treeline after a doubling of atmospheric CO_2. Redrawn from Figure 7 of GCASR (1995).

especially where it is thin, can undergo thawing from the bottom and sides as well as the top. Farther north, where the permafrost is coldest and thickest, most of the thawing will be at the top of the layer, and there the portion thawed may become incorporated into the active layer. In northern Alaska where the permafrost is near 600 m thick, thawing at the bottom of the layer during the warming of the past 100 years is estimated to be only approximately 0.8 m.[14]

Among the most visible consequences of the decades of recent warming are the caved-in shorelines and submerged trees along thermokarst lakes in the discontinuous permafrost zone, see Figure 4.45/Plate 18. These are particularly noticeable in the Mentasta Pass area in Alaska and along the Alaska Highway in Yukon Territory near the Alaska border in Canada.

Somewhat less obvious is the deterioration of roadways and building foundations in the discontinuous permafrost zone, less obvious perhaps

---

14. Lachenbruch et al. (1982).

because it is masked by the deterioration from alteration of the ground temperature regime caused by the roadways or the buildings. In Alaska, the region around Gulkana in the Copper River basin (where fine-grained soil abounds) is a prime example of a region where thawing permafrost is creating rollercoaster roads and tilting buildings. Reports indicate that virtually every building in this area has foundation problems. Continuing warming should lead to a slow northward shift in the zone where the deterioration is most serious, leaving at its southern edge a zone where stabilization will eventually occur. Hence areas such as that around Gulkana may eventually benefit—at the expense of worsening conditions for neighbors to the north.

One potentially major consequence of permafrost thawing with climatic warming is the release of substances formed from biological processes and then locked into the permafrost during times long past. These substances include carbon dioxide, methane, nitrates, and phosphates. Plant growth pulls carbon dioxide from the atmosphere and introduces part of it into the soil. Anaerobic bacterial action decomposes humus matter to form methane, some of which is retained in the soil. Bacterial action pulls nitrogen from the atmosphere and converts some of it to nitrates that rainwater or meltwater leach down into the soil. Phosphates derive from weathering of phosphorous-rich rocks, are taken up by plants and then reintroduced into the soil by water.

Recently reported studies of boreal forests provide clear evidence that carbon dioxide locked into permafrost several hundred to 7000 years ago is now being given off to the atmosphere as warming climate melts the permafrost. Investigation of a black spruce forest showed that most of the old carbon dioxide given off was coming from a zone 40 to 80 cm deep, well beneath the biologically active layer containing moss and tree roots.[15] The authors of the study note that this relict carbon dioxide represents a massive source since it is estimated that the carbon dioxide contained in the seasonally and perennially frozen soils of boreal forests is to 200 to 500 billion metric tons, enough if all released to increase the atmosphere's concentration of carbon dioxide by 50%. Hence, it is possible that the release of carbon dioxide from melting permafrost during warming, or locking it into newly frozen soil during cooling, may accelerate climate change. Methane released by thawing soil into the atmosphere, absorbs sunlight, like carbon dioxide, but its concentration in the atmosphere is low, and so its overall effect may be minor. For that matter, the uptake or

---

15. Goulden et al. (1998).

release of carbon dioxide due to changing volumes of frozen soil may also be minor compared to changes in the concentration of atmospheric carbon due to other causes.

It appears that nitrates and phosphates released by thawing permafrost in the Fairbanks, Alaska, area and perhaps elsewhere may be the cause of high levels of these substances in water drawn from some wells.[16] In well water from the hilly area northwest of Fairbanks, one investigator has found concentrations of nitrate that exceed allowed federal standards by up to 500%,[17] and some of my neighbors and I in the same area are having trouble maintaining proper chemistry in our hot tubs because of extremely high phosphate levels. Owing to the sparsity and location of houses in the area it is virtually certain that the phosphate is from a natural source.

Another consequence of thawing permafrost and increasing thickness of the active layer is release of frozen peat and other plant material that serves as food for bacteria, aquatic insect larvae, and other invertebrates which in turn feed fishes and birds. Studies of them indicate that the fossil food locked into permafrost in northern Alaska accounts for a substantial portion of their energy intake during the thawing season.[18] Thus it is possible that changes in the availability of the fossil plant material as permafrost thaws or increases in volume could modify habitat enough to affect bird migration and survival.

▨　▨　▨

The current trend toward warming and thawing of permafrost is creating topographic changes both subtle and obvious, and these changes are occurring rapidly, at least in terms of geologic time. Yet they are slow enough in human time that people can be unaware of them unless they pay special attention. Development of thermokarst topography and alteration of drainage patterns are likely changes that could affect human habitation and engineering projects, but these are hard to predict, and so perhaps the best approach is just to be aware that they may happen.

One possibility is that the current warming trend will reverse at any time because of natural climatic cycling, thereby causing resurgent permafrost to become extensive enough to cause problems in some areas. An example of that is the demise of the Norse colonies established in Greenland

16. Robert. B. Forbes, personal communication, (1998).
17. Johnson-McNichols (1987).
18. Schell (1989) pp 159–60.

during the tenth century when the climate was much warmer than it is now. Dependent on agriculture, the colonies flourished at first but then the Little Ice Age began. By 1400 A.D. enough permafrost had formed in farming plots to contribute to the abandonment of the colonies.[19]

▪ Chapter Summary

The earth's climate is continually changing due to variations in the sun's energy output, cyclic variations in sun-earth geometry, and changes occurring within the earth system, including its atmosphere. The human race is producing part of the change to the earth system. Increasing solar output and human activities appear to be about equally responsible for the warming trend now underway. What the future will bring is unknown, other than that more change is assured. Although some is still forming, much of the permafrost now in the ground has persisted for thousands of years, and many of the cryogenic landforms now in evidence developed long ago. During a warming trend like that underway permafrost becomes less widespread and it becomes weaker, less able to bear up under load. The result, especially in the discontinuous permafrost zone, is the development of thermokarst topography, increasing problems with building foundations and accelerated disintegration of roadways. Also, gases such as carbon dioxide and chemicals such as nitrates and phosphates that were locked into the frozen ground long ago are released, perhaps accelerating climate change or causing environmental problems.

---

19. Brooks (1973); Pringle (1997).

Appendix A

Additional Background and Some Fundamentals

▪ Amount of Water on the Earth

Water is the most abundant single chemical substance near the surface of the earth, and it is far and away the most plentiful of the two liquids occurring there in million-ton or greater quantities. Liquid petroleum is the other, but water is a billion-billion times more prevalent.

The mass of water in the hydrosphere (by definition, the waters covering the earth's surface, mainly within the oceans) is 1.66×10^{24} grams. Since the earth has a surface area of 5.1×10^{18} cm^2, this is enough water to cover it with a uniform layer 3 km (2 mi) deep.

The earth's lithosphere (the outer part of the solid earth composed of rock like that found at the surface, a layer with mean thickness 35 km beneath the continents and 5 km beneath the oceans) is estimated to contain 2 to 3×10^{24} grams of water. Since the lithosphere itself has mass 2.4×10^{25} grams, it is roughly 10% water.

Even more water is contained in reserve within the next layer down, the mantle, the solid part of the earth directly below the lithosphere. It is about 3,000 km thick and rests directly above the earth's core. The reserve water in the mantle amounts to 2×10^{26} grams, 10 times as much as resides in the hydrosphere or the lithosphere.

The atmosphere also contains water, but the total amount is far less than in the hydrosphere or lithosphere. The entire mass of the atmosphere is 5×10^{21} grams, about 1/300 that of the hydrosphere, and only a small portion of the atmosphere is water. The atmospheric water content ranges from a tiny fraction of 1% in the cold arctic regions to roughly 5% in the warm tropics.

▪ Atoms, Molecules and the Forces That Hold Them Together

Water molecules and the molecules of all other substances are made up of atoms, and atoms are made of electrons, protons, and neutrons (in turn thought to be composed of even more fundamental particles, quarks and gluons). When electrons, protons, and neutrons come into proximity they join together in certain set ways determined by whatever forces come into play. Those forces, in turn, are determined by the characteristics of the particles. Very strong (so-called nuclear) forces between the protons and neutrons bind these particles together so tightly that they form into a compact entity called the atomic nucleus. Electrons bound far more loosely to the atomic nuclei complete the structure of atoms. The binding forces here are weaker than those within the nuclei and are of different type. Their action permits each atom's electrons to stand far out away from its nucleus—much as the earth stands out away from the sun, and as the moon stands out from the earth. **Table A1** gives specific information on the characteristics of electrons, protons, and neutrons. Notice that:

1. The mass of the proton is 1,836 times the mass of the electron and is slightly less than the mass of the neutron. Consequently, virtually the entire mass of an atom is contained within the atomic nucleus.

2. All three elementary particles are very tiny, but roughly of the same size: 1 to 10 x 10^{-13} cm in diameter. By contrast, the diameter of an atom is roughly 1 x 10^{-8} cm. This atomic diameter is 10,000 to 100,000 times larger than the diameters of the electrons, protons, and neutrons, so atoms—like the solar system—are mostly empty space.

Table A.1 The Basic Building Blocks: Electrons, Protons, and Neutrons Compared

QUANTITY or CHARACTER	ELECTRON	PROTON	NEUTRON
MASS (in mass units = 1.66×10^{-24} g)	1/1836.2	0.9	1.009
DIAMETER (in cm)	$1\text{--}10 \times 10^{-13}$	$1\text{--}10 \times 10^{-13}$	$1\text{--}10 \times 10^{-13}$
CHARGE (in elementary units = 4.8×10^{-10} esu)	−1	+1	0
SPIN (angular momentum) times $2\pi/h$	+ or $-\frac{1}{2}$	+ or $-\frac{1}{2}$	+ or $-\frac{1}{2}$
MAGNETIC MOMENT (in nuclear magnetons)	−1,840 (directed opposite to spin)	2.79 (directed with spin)	−1.9 (directed opposite to spin)

3. Each electron carries an elementary unit of negative electrical charge, and the proton a positive charge of exactly equal magnitude. The neutron is neutral; that is, it carries no electrical charge.

4. All three elementary particles spin like tops on their own axes and therefore have angular momentum, usually referred to as spin, which by convention is angular momentum multiplied by $2\pi/h$, where h is Planck's constant, a universal constant of nature that relates the energy of a quantum of radiation to the frequency of the oscillator that emits it. The spins are equal in magnitude.

5. A consequence of the spin is that electrons, protons, and neutrons each have magnetic moments; that is, they behave like little bar magnets. The magnetic moments contribute to the magnetism, and since the magnetic moment of the electron is so large, the magnetic character of atoms and molecules is almost entirely determined by their electron configurations.

About Force in General

A force is anything that changes the state of rest or motion of matter. The force F required to produce acceleration a in a mass m is $F = ma$. In everyday experience only the gravity force and the electromagnetic force come into play. The gravity force affects mainly massive objects like rocks, people, planets, and stars, but it has comparatively little effect on tiny objects like electrons, protons and neutrons. On that scale, electromagnetic force is much stronger than the gravity force so electromagnetic force is the one that binds atoms and molecules together.

In discussing forces, people often use the terms "force field" or simply "field" to describe how strongly a force will act on a test object placed at various places in the region where the field is effective. Thus, we use such terms as "gravity field," "electric field," and "magnetic field." In many instances it is important to specify the direction of a field as well as its magnitude. A convenient way to do this is with vectors, quantities having both magnitude and direction.

Nuclear Forces

In addition to gravitational and electromagnetic forces, two other classes of forces are recognized. Known as the nuclear strong interaction forces and the nuclear weak interaction forces, these somewhat mysterious, poorly understood forces are the powerful glues that hold nuclei together. They act only on distance scales comparable to that of atomic nuclei (10^{-12} cm) and so do not enter into the bonding between atomic nuclei and the electrons in the outer part of an atom nor into the bonding together of atoms to create molecules.

The Gravity Force

Two masses m_1 and m_2 separated by distance r attract each other with a force $F = Gm_1m_2/r^2$ that is directed along the imaginary line joining them, where G is a constant called the universal constant of gravity (see Figure 2.2). The gravity force causes the earth to cling to the sun, and the moon to the earth, but the gravitational forces between electrons and atomic nuclei are too weak to be of consequence. However, of all the known forces, gravity is the simplest and most familiar.

Electromagnetic Force

Electromagnetic force operates on matter through its electrical characteristics. Electromagnetic force has two parts. One part (the electro part) depends on the *position* of electrical charge, and the other part (the magnetic part) depends on the *motion* of charge.

Electric Force

The part of electromagnetic force depending only on the position of charge is called the coulomb force or electrostatic force. The coulomb force F between two electric charges q_1 and q_2 separated by distance r in a vacuum is F $= q_1q_2/r^2$. Like the force of gravity, the coulomb force acts only along the imaginary line drawn between the two objects involved and it is inversely proportional to the square of the distance between them. But the coulomb force depends on charge instead of mass, and charge can be negative or positive. The coulomb force between like charges is repulsive and between charges of opposite sign it is attractive.

Magnetic Force

Magnetic force is itself the product of moving charge, and it acts only on moving charge. When a charged particle is in motion relative to a reference frame, it generates a force that can act on another charged particle in motion relative to the same reference frame. Unless the second charge is in motion relative to the reference frame the force generated by the other particle does not act on it, nor does it act if the particle's direction of motion happens to be exactly in a direction at right angles to the direction of motion of the initial charge. If the second charge does experience a force, that force will try to push the charge in a direction at right angles to the direction it is moving. This obviously is a complex situation. It is like saying that a tennis ball moving across a court will exert a force on another tennis ball only if the second tennis ball is also moving across the court, and in a direction not at right angles to the motion of the first tennis ball—and that the force acting will try to make the second tennis ball turn aside from the direction it is moving. The complexity of the situation is reduced somewhat by bringing in the concepts of electric current and magnetic field. Motion of charge, be it a single charge or a collection of charges moving in the same direction, is defined as electric current, and an electric current generates a force field called a magnetic field. The magnetic field has a direction at right angles to the motion of its generating current (again see Figure 2.2). A magnetic field affects only moving charge, and it acts to push the charge in a direction that is at right angles both to the direction of the magnetic field and the direction the charge is moving. Magnetic materials—those that generate magnetic fields and are affected by them, for example, a simple bar magnet—have this property owing to motions of electrons within the atoms of the materials. In each atom electrons move in looping patterns about the atomic nucleus and so that motion gives rise to magnetic field, but the primary motion that generates the magnetic character of materials is the spin of individual electrons on their own axes. As shown in Table A1, electrons have very large magnetic moments, which makes them highly susceptible to magnetic force as well as highly effective generators of magnetic force. The shape of the magnetic field of an electron spinning on its own axis is that of a bar magnet, and that is the same shape as the field generated by an electron moving in a circular pattern to form a current loop. Called a dipole field, this is the simplest magnetic field known.

The magnetic forces produced by rotating electrons are crucial elements in the structuring of atoms and molecules because when two electrons come close together their magnetic fields interact to cause either strong repulsion or strong attraction, according to how they are oriented relative to each other. The interaction is similar to that between two bar magnets brought close together, but more complex because the repulsive electrostatic coulomb force is acting too.

The Assembly of Atoms

Recall the Universal System Happiness Rule presented in Chapter 2: every system is happiest when it contains the least amount of free energy, or more precisely, **The stable, equilibrium state of a system is the state of minimum free energy**. One of the places this rule comes into play is in the assembly of all the known elements from the basic building blocks: the atomic nuclei (containing protons and neutrons) and electrons. Electromagnetic forces operating between atomic nuclei and electrons cause these entities to join together to form atoms only when the joining reduces the overall free energy that the nuclei and electrons previously possessed. Furthermore, when these fundamental building blocks do join together to form the more than 100 known elements, they always enter a configuration structured to have the least possible amount of free energy.

The methodology used to calculate how much free energy a particular configuration of atomic nuclei and its surrounding electron cloud will have is called quantum mechanics. The rules of quantum mechanics also dictate which configurations are stable enough to stay around long enough to be noticed.

Owing to the Universal System Happiness Rule which governs all things, any electrons that join with an atomic nucleus to form a stable atom will position themselves in a configuration that gives a state of minimum free energy. The rules of quantum mechanics merely specify for each atom the allowed electron configurations and a means of calculating the free energy associated with each possibility. The possibilities are called states,[1] and the formulas for calculating the energy associated with each state are called state functions. Each state function incorporates four variables called quantum numbers. They are designated as n, l, m, and m_s, and certain strictures govern them. One requirement (the Pauli exclusion principle) is that no two electrons in an atom can have the exact same set of quantum numbers. Another is that the first three (n, l, m) must be whole numbers (integers), and they must obey the inequality relation:

$$n > l \geq |m|$$

The fourth quantum number, m_s, can take on only the value $+\frac{1}{2}$ or $-\frac{1}{2}$. Each of the four quantum numbers contributes to specification of the energy carried by an electron, but n, called the total quantum number, is the most important. This quantum number relates directly to the distance of an electron out from the nucleus, l has to do with the vigor of rotation of an electron about the nucleus, m specifies the orientation of the orbiting electron, and m_s relates to the spinning of an electron on its own axis, whether the spin is in one direction or the other.

In essence, quantum mechanics is a computer software program that lets a person see how to construct all the atoms in the periodic table. Those are all the atoms of the universe, and some of them, especially the hydrogen and oxygen atoms that combine to form water, determine the intricacies of permafrost.

Given an atomic nucleus to start with, plus the one or more electrons needed to combine with it to create a neutral atom, the first step is a computation of the state function to determine the configuration having the least amount of free energy so that the Universal System Happiness Rule is satisfied. The simplest possible nucleus is that of hydrogen, consisting of only one proton. The calculation of the state function shows that the configuration of least free energy occurs when $n = 1$ and, because of the inequality relation above, both l and m must equal zero. If a hydrogen atom is in complete isolation the energy carried by its electron might not depend on m_s, but if electromagnetic fields or other atoms are present, the least free energy (the greater stability) may be associated with one or the other of the two allowed values of m_s.

---

1. Not to be confused with states of matter, i.e., solid, liquid, or gaseous. The meaning of "state" here is much as in the formal statement of the Universal System Happiness Rule, "the stable, equilibrium state of a system...," and it specifically relates to the energy an electron possesses.

Helium, with two protons in its nucleus, is next in line. The calculated state function for the configuration of the two electrons needed to make the atom electrically neutral shows that the least free energy occurs with one electron in the state $n = 1$, $m_s = +\frac{1}{2}$ and the other in the state $n = 1$, $m_s = -\frac{1}{2}$. Again, both l and m must equal zero. The configuration is spherically symmetrical (also called closed-shell), and such a configuration is known to interact little or not at all with other atoms or molecules. Thus, helium is inert.

Lithium, with atomic number $Z = 3$ (the number of protons in the nucleus) requires three electrons, but only two slots with $n = 1$ are available and the third must have a principal quantum number greater than 1. The configuration with least free energy is ($n = 2$, $l = 0$, $m = 0$, $m_s = +$ or $-\frac{1}{2}$). And so it goes on up through all the known elements, more than 100. To form the next element, the principle quantum number n must be jumped by 1 whenever enough electrons have been added to form the symmetrical, closed-shell configuration. Those configurations occur at $Z = 2$ (helium), $Z = 10$ (neon), $Z = 18$ (argon), $Z = 36$ (krypton), $Z = 54$ (xenon) and $Z = 86$ (radon). The closed-shell configuration of these elements has much lower free energy than other electronic configurations so they refuse to combine with other elements. The other elements also seek to develop the closed-shell configuration by either giving up electrons or snatching them away from atoms. That is why most atoms like to combine chemically with others to create molecules: by so doing they approach the minimal free energy associated with the highly stable, closed-shell configuration. An example—one particularly relevant to the topic of this book—is the water molecule. By combining with two hydrogen atoms, an oxygen atom approaches the minimum-energy, stable closed-shell configuration of the neon atom, and each hydrogen atom in the water molecule approaches the minimum-energy, stable configuration of the helium atom.

▪ The Bonding of Atoms into Molecules

The electromagnetic forces that bind electrons to atomic nuclei to create atoms also act to bring atoms together to form molecules. The combination of an atom's nuclear size and the configuration of its electrons determines the relative importance of the electrostatic coulomb and magnetic forces to the molecular bonding. Although both electrostatic and magnetic forces play a part in bonding atoms together, some circumstances favor strong electrostatic force (called ionic or valence) bonding, while others favor magnetic force (called covalent) bonding.

Ionic (electrostatic, also called valence) Bonding

An atom such as hydrogen, lithium, sodium, or potassium that has just one electron in its outer shell maintains only a loose grasp on this electron and would happily get rid of it in order to achieve the stability of the closed-shell configuration. Similarly, and for the same reason, atoms such as fluorine, chlorine, and bromine that are shy one electron of having a closed-shell configuration are likely to steal away an electron.[2] The molecule in ordinary salt, NaCl, is a typical example of a successful theft that makes both sodium and chlorine atoms very happy (in the metaphoric sense of the Universal System Happiness Rule). Having given up an electron, the sodium atom obtains the closed-shell configuration, as has the chlorine atom by receiving one. The overall system, the NaCl molecule, has less free energy than did the independent Na and Cl atoms, hence more stability.

However, the thievery has created a new imbalance since the Na atom has one more proton than its number of electrons, resulting in a net positive charge on that atom. Similarly, the Cl atom has an extra negative charge. The electrostatic attraction between the excess charges, called ionic bonding, binds the two atoms together. Ionic bonding is strong, as is covalent bonding, another kind of molecular glue that depends on magnetic forces.

Covalent Bonding

Atoms constructed such that they can join with other atoms in a manner that allows each of them to approach the stable closed-shell configuration by sharing two electrons more or less equally between them are said to undergo covalent bonding. The hydrogen molecule is the simplest example of pure covalent bonding; by equally sharing an electron pair each atom feels it has achieved the stable closed-shell configuration of the inert helium atom.

---

2. Recall that the bonding in the water molecule is about 40% ionic as well, so oxygen behaves somewhat like the highly reactive elements lacking but one electron to have the closed-shell configuration.

Quantum mechanical calculations show that as two distant hydrogen atoms are moved together they at first show a mutual repulsion but then as the distance decreases a strong attraction develops, and the atoms can join to form the hydrogen molecule H_2. For that to happen, the two electrons involved must have opposite spin because the chief glue bonding the hydrogen atoms together is the mutual magnetic attraction between oppositely spinning electrons. They cling together in what is called an electron pair. (Recall from Chapter 2 the analogy with the team of oxen.)

Exactly what is happening when electrons pair up to bind molecules together is a bit hard to envision because electrons are not really the little points or balls of charge that we usually think of when we ponder them and their interactions. A more realistic view is to think of an electron as a smeared-out region of charge that can both rotate on its own axis (the spin) and also orbit about one or more nuclei in some fashion. In the hydrogen atom the electron is a spherical smear of charge most dense about 0.53×10^{-8} cm out from the proton nucleus, but in the hydrogen molecule both electrons form an elongated smear of charge that encompasses both nuclei but which is most dense in the region between the nuclei.

Linus Pauling, Nobel Prize-winning expert on chemical bonding, suggested that a hydrogen molecule might be thought of as something like two steel balls (the atomic nuclei) vulcanized into a tough piece of rubber (the electron pair). Another analogy to the effects of electron pairing makes use of two annular (ring) magnets. This is a good analogy in several ways because the magnetic field of each magnet is exactly that of a spinning electron, and the effects can be seen and felt as well as thought about. If one magnet is brought close to the other, both will try to orient themselves in the same plane, and when they are in that plane they will either strongly repel (as do two electrons having the same spin) or strongly attract (as do two electrons of opposite spin). It is, in essence, the Pauli exclusion principle at work: the two magnets (electrons) can join if their axes (spins) are exactly opposite, and it is only the relative orientation that matters, not the orientation relative to some fixed coordinate system. Seeing the magnets flip to orient themselves properly and feeling the strength of the attractive force makes it easy to think about the effectiveness of the covalent bond. Due to the strong effects of electron pairing, covalent bonding is so prevalent that is sometimes called *the* chemical bond.[3] In the water molecule, approximately 60% of the bonding between oxygen and hydrogen atoms is covalent, and the other 40% is ionic.

---

3. Pauling (1970) p 148.

The Hydrogen Bond

In addition to the strong covalent and the strong ionic bond, another much weaker but important linkage occurs with hydrogen atoms because of their unique structure, a single electron encompassing a small nucleus composed only of one proton. This linkage, called the hydrogen bond, is only about 4% as strong as a covalent bond, but it has extremely important consequences for permafrost and other matters of interest.

Atoms such as lithium, sodium, and potassium possessing a single electron in their outer shells have one or more underlying closed electron shells to act as an electrostatic shield for the nucleus when the outer electron covalently locks into another atom. By contrast, the hydrogen atom lacks an underlying shell, so when a hydrogen atom combines covalently with another atom the hydrogen nucleus is left hanging out nakedly. The only thing cloaking it from the rest of the universe is the pair of electrons shared with the other atom, and most of the charge swarm comprising that pair lies between the two nuclei. Thus the hydrogen nucleus sticks out like a sore thumb, and because it carries positive charge it is capable of electrostatically attracting the electrons of other nearby atoms strongly enough to form a fairly stable linkage known as the hydrogen bond. For that to happen, the nearly naked proton must be able to come in between two electron pairs owned by another atom and lock electrostatically to them. Only small atoms such as nitrogen, oxygen, and fluorine participate in hydrogen bonding because lack of space prevents larger molecules from coming close enough to form stable bonds.

The Van der Waals Attraction

When two molecules come into proximity a feeble electrostatic force arises from the fact that the mutual attraction between the (positively charged) nuclear portion of one molecule and the (negatively charged) electrons of another is slightly greater than the repulsion of its electrons from the other's electrons and the mutual repulsion between the two nuclear entities. The net result is the Van der Waals attraction, a force so weak that it almost fails to qualify as a type of chemical bonding. The larger the molecules involved, the greater the Van der Waals attraction, and it is sufficient in some instances to cause a weak bonding between molecules. One example is O_4, an insignificant and fragile molecule formed of two molecules of O_2 by the Van der Waals attraction. While not important in chemical bonding, the Van der Waals attraction is a significant force between molecules when close together, as in a liquid or solid.

When a liquid is heated it is the Van der Waals attraction that must be overcome by thermal agitation to cause boiling. In general, because the Van der Waals attraction between heavy molecules is greater than that between light molecules, substances with higher molecular weights tend to have the higher boiling points.

Polar and Nonpolar Molecules; the Molecular Dipole Moment

Molecules differ in their electrical behavior. Those that are highly symmetric tend to exhibit no outward electrical behavior; that is, they tend neither to create electric fields nor to be affected by them. They are electrically neutral because at any point in space near them the electric fields due to the positive charges in the nuclei just balance the electric fields due to the electrons. Such molecules are called *nonpolar.* Contrasting with them are *polar* molecules, those having such a lack of symmetry that the distribution of electrical charges in the molecules produces a net electrical field at any nearby point.

Covalent bonding promotes electrical symmetry, so molecules with strong covalent bonding tend to be symmetrical, and hence nonpolar. Ionic bonding, by contrast, tends to stretch the molecular charge distributions and thereby form the electrically active polar molecules. The bonding between the atoms of any molecule always involves some covalent bonding and some ionic bonding, so every molecule falls somewhere in a range that extends from strictly nonpolar to highly polar. Certain trends pertain, and it is possible to describe quantitatively the degree of polarity. One trend is that those elements located nearest to the center of the periodic table tend most toward covalent bonding and those near the right-hand and left-hand sides tend toward greater ionic bonding. Also, the bonding between atoms of similar size trends toward covalent bonding (as in $H + H \rightarrow H_2$) and those of different size trend more toward ionic bonding (as in $H + Cl \rightarrow HCl$). When two unequal atoms combine covalently, the newly formed electron pair is centered neither midway between the two atomic nuclei nor entirely within the new closed outer shell of the atom having the stronger attraction for the pair. The consequence is a stronger ionic bonding force (electrostatic) than would occur were the atoms of equal size.

Figure A.1 shows how the lack of symmetry comes about when the two grossly unequal atoms hydrogen and chlorine join to form the hydrogen

WATER MOLECULE

HYDROGEN CHLORIDE MOLECULE

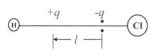

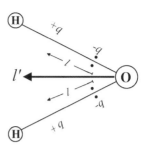

Figure A.1 The relatively large chlorine atom pulls the electron pair away from the light hydrogen atom, creating an electrical imbalance that can be described as though charges $+q$ and $-q$ are located on the joining axis a distance l apart. This situation creates a dipole moment $= ql$. The large dipole moment of the water molecule is the sum of the two projections of the dipoles between the oxygen atom and the hydrogen atoms onto the one line of symmetry of the water molecule. Diagram compiled following Figure 5.2 of Deming (1975).

chloride molecule. The heavy chlorine atom pulls the electron pair away from the light hydrogen nucleus so that more negative electrical charge lies toward the chlorine end of the molecule. An equivalent statement is that more positive charge lies toward the hydrogen end of the molecule, but the situation is most usefully described by using a bit of each of the equivalent statements; that a small positive charge $+q$ is toward the hydrogen end and an equal negative charge $-q$ lies toward the chlorine end and the two are separated by distance l. Such a configuration with two opposite charges is called an electric dipole, and it has defined dipole moment ql.

This is a highly significant matter because if molecules having dipole moments are placed in an electric field, the field will try to orient them along the direction of the field, and the resulting distribution of charge will act to decrease the magnitude of the field. The force that the electric field exerts to turn the dipoles is proportional to the length of the dipole arm l. Nonpolar molecules are those with very small or zero l, and polar molecules are those with the larger values of l. A practical way to measure the polarity of a substance is to measure its dielectric constant (ε). The dielectric constant is proportional to l, and it describes the extent to which an imposed electric field will be decreased by the orientation of the dipoles. Thus it is possible to construct an electrical apparatus to measure ε. The dielectric constant of a vacuum is 1, and its value for nonpolar substances typically ranges between 1 and 2. The dielectric constant for strongly polar substances ranges upwards of 10. Water has $\varepsilon = 78.5$, so it is highly polar. Molecules with more than one axis of symmetry always are essentially non-polar. Examples are: hydrogen, H_2 [$\varepsilon = 1.23$]; nitrogen, N_2 [1.45]; oxygen, O_2 [1.51]; carbon tetrachloride, CCl_4 [2.24]; silicon tetrachloride, $SiCl_4$ [2.4]; carbon dioxide, CO_2 [1.60]; and chlorine, Cl_2 [2.10]. Molecules with only one axis of symmetry are essentially polar. Examples of highly asymmetric and therefore strongly polar molecules are: water, H_2O [$\varepsilon = 78.5$]; ammonia, NH_3 [17]; silicon dioxide, SiO_2 [15]; hydrocyanic acid, HOCN [118]; and nitrogen fluoride, NF_3 [84].

■ Supercooling and the Critical Radius

When water begins to convert to ice, the overall system changes from being one of only liquid water to one consisting partly of water, partly of ice, and partly of the interface between the two. This interface has free energy associated with it in an amount that depends on the area and shape of the interface. If the interface is very small and has high curvature (as would be true if the ice were in the shape of a tiny sphere) any growth of the interface causes the overall system free energy to increase, so conversion of water to ice does not occur. Yet if the temperature drops low enough, or if a foreign object with radius bigger than what is called the critical radius is introduced into the liquid, freezing can occur without increasing the system's overall free energy. Then, in fact, freezing decreases the system's overall free energy. The mathematics of the situation is as follows:

The free energy of a system G_{total} is the sum of the free energies of its parts. Assume a system composed of a volume of water V in which is immersed a tiny ice crystal of radius r, which will have volume $4/3\pi r^3$ and surface area $4\pi r^2$. This system has three parts, the water, the ice, and the interface between. The total free energy is

$$G_{total} = G_{water} + G_{ice} + G_{interface}$$
$$= g_{water}(V - 4/3\pi r^3) + g_{ice}(4/3\pi r^3) + \sigma_{iw}(4\pi r^2),$$

where g_{water} and g_{ice} are the free energies per unit volume and σ_{iw} is the interfacial tension between ice and water, which is interpreted to be the work required to create a unit area of new interface. If the ice crystal grows by increasing its radius an amount Δr then the total free energy changes to

$$G + \Delta G = g_{water}[(V - 4/3\pi(r + \Delta r)^3] + g_{ice}[4/3\pi(r + \Delta r)^3] + 4(\sigma_{iw}(r + \pi r)^2.$$

Subtracting the first equation from this one gives

$$\Delta G = 4/3\pi g_{water}[(r + \Delta r)^3 - r^3] + [4/3\pi g_{ice}[(r + \Delta r)^3 - r^3]$$
$$+ 4(\sigma_{iw}[(r + \Delta r)^2 - r^2].$$

This equation expresses how the free energy of the system changes in terms of the change in radius of a growing ice crystal. To determine ΔG, the free energy associated with the formation of an ice crystal of radius Δr, we set $r = 0$, and obtain

$$\Delta G = 4/3\pi \Delta r^3 (g_{ice} - g_{water}) + 4\pi \sigma_{iw} \Delta r^2$$

$\Delta G = 0$ at $\Delta r = 0$, and if g_{ice} is smaller than g_{water}, ΔG again equals zero when $\Delta r = 3\sigma_{iw}/(g_{water} - g_{ice})$. At any temperature below its freezing point the free energy of water is greater than the free energy of ice, so the plot of energy ΔG against Δr has two zero values, as shown in **Figure A.2**. Thus the growth of an ice crystal starting from scratch involves an increase in free energy, and this increase is the reason why water and other liquids supercool. The increase continues until

$$\Delta r = 2\sigma_{iw}/(g_{ice} - g_{water}).$$

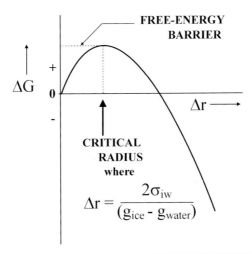

Figure A.2 At the critical radius Δr, the ice-water system can either increase or decrease its amount of ice since either direction is a move in the direction of lower free energy. Based on Figure 3-4 of Knight (1967).

At this value of the radius, called the critical radius, either melting or growth causes a decrease in system free energy so, in principle, either can occur. Typically, the critical radius is approximately 10^{-7} cm, several times the dimension of the molecules in the forming crystal.[4] The value of ΔG at the critical radius is called the *free-energy barrier* or *activation energy of nucleation*. That is the barrier water molecules must leap in order to convert to ice. If crystals of greater radius contact the liquid water, the conversion to ice is spontaneous, and supercooling terminates.

▪ How Water's High Viscosity, High Surface Tension, and Wetting Ability Affect Its Behavior in Capillaries

All gases and liquids tend to resist changes to their form. Called viscosity, this resistance is a sort of internal friction that increases according to the degree with which the molecules of a substance cling to each other. Water has higher viscosity than almost any other liquid; the reason is the high degree of association of water molecules caused by hydrogen bonding. The viscosity of water changes considerably with temperature because the warmer the water, the fewer the associated molecules. A decrease in temperature from 100°C to 90°C increases water's viscosity by 12%; an equal change from 10°C to 0°C increases the viscosity by 37%, and the viscosity increases an additional 42% as water is supercooled from 0°C to –10°C. Supercooled water at –10°C is 9 times more viscous than water at 100°C.

A related characteristic is water's remarkably high surface tension, the tendency of the surface of a liquid to contract and act like a stretched membrane. Surface tension is due to the attractive forces between those molecules of a substance that lie at its surface and those immediately below. These forces are no different from the attractive forces between molecules within a substance; it is just that right at the surface those forces operate only to the interior side. For example, inside a volume of water any molecule or group of associated molecules is subject to attractive forces pulling from all directions, but if a cut were made through the volume and everything to one side removed, the forces pulling to that direction would suddenly be absent. Subject only to forces on one side, molecules at the surface would then try to draw together to achieve new

---

4. Knight (1967) p 32.

equilibrium positions, and the change in position would increase the potential energy (free energy) of the surface molecules.

But, as always, the equilibrium position is that involving minimum free energy, and because those molecules at the surface have higher free energy than those interior, the system will try to minimize its surface area.[5] Since a sphere has least surface area per unit volume, the surface will develop curvature in its attempt to achieve this configuration. The higher the surface tension and the smaller the volume, the more likely is the substance to attain spherical shape. Surface tension does not increase the pressure within a system by forcing a contraction. Instead, the downward forces on the surface molecules pull them down out of the surface, and as this happens these molecules lose potential energy (free energy). Notice that the subtraction of these molecules from the surface must make the surface smaller if the substance is to stay together. It is useful to remember that surface tension really is the free energy associated with the surface, and that free energy directly depends on the sum of all the potentials arising from mutual interaction between pairs of particles totally within the liquid plus interactions between the liquid's surface particles and those of the solid or fluid with which they are in contact.

Because of the hydrogen bonding that is responsible for making it an associated liquid, water has the highest surface tension of all commonly occurring liquid molecular substances. Mercury, also a commonly found liquid but not a molecular substance, has even higher surface tension. Mercury's surface tension is approximately 470 dynes/cm, water's is 73 and most other liquids have surface tension in the range 10 to 40 dynes/cm. (Some organic liquids with hydrogen and oxygen in their molecules have relatively high surface tensions also, as does hydrogen peroxide (H_2O_2) with a surface tension of 92.)

The effect of high surface tension is readily observed by putting a small quantity of mercury on a horizontal glass plate. The mercury forms a sphere, and if the quantity is increased, the mercury will spread out into a flat-topped blob with rounded edges. Water similarly placed on the glass plate does not bead up nearly as much as mercury does. Of course water has far lower surface tension, but another important characteristic of water comes into play. This is water's ability to wet (adsorb to) glass or almost any substance that contains oxygen because of water's hydrogen

---

5. The free energy—equivalent to potential energy—is higher because work is done on the surface molecules to bring them to their positions from the positions they would have occupied had the surface not been there, that is, had they been interior molecules.

bonding ability.[6] The hydrogen bond reaches out for the oxygen atoms contained in glass (SiO_2) or any other substance such as cotton fiber, rock, clay, and any organic and inorganic soil particles. (Among the substances that water does not wet are the paraffins, the open-chain hydrogen and carbon molecules [C_nH_{2n+2}] that form the bulk of crude petroleum molecules: methane, ethane, propane, butane, octane, etc.) Mercury does not wet glass, so when placed in a small-diameter glass tube (a capillary tube) it forms a curved surface that is convex, whereas water's meniscus is concave.

The comparison between the wetting abilities of water and of mercury points out that a substance's surface behavior depends on what is located adjacent to its surface. Substances such as soap or oil in contact with water lower its surface tension. (A good soap or detergent contains so-called double-hook molecules, ones that on one end tend to attach to water, and on the other end tend to attach to the particles of dirt or other substance that is to be removed from whatever is being cleaned. In this fashion, soaps and detergents act more mechanically than chemically.)

Of geologic importance is that water's high surface tension makes its raindrops like bullets: they can shatter minute fragments off the hardest of rocks. Water also is an effective eroding agent because its hydrogen bonds allow it to wet soils. When a raindrop falls and gouges a dent it also shatters; then, as new water spheres form from the fragments, they scoop up small bits of dirt. The resulting suspended fine material tends to increase the proportionate amount of runoff from a bare soil by clogging up the soil's pores. On the other hand, the wetting contributes to retention of water by soil and the combination of high viscosity and high surface tension tends to slow runoff in rivers.

Wetting ability refers to the degree of adsorption undergone by a liquid. Adsorption (not to be confused with absorption, which is the process of one substance taking another into its internal structure) is the attaching of the molecules of a substance (the adsorbate) to the walls of another (the adsorbent). The adsorbent's walls might be perfectly smooth or they might contain nooks and crannies or even channels than run through the substance. Because of their ability to form hydrogen bonds with soil, protein, and other materials (even glass), water molecules wet, i.e. adsorb[7] easily to, these materials. Also, water's highly dipolar nature causes it to

---

6. Silicones are exceptions because their organic molecular structures lock in the oxygen molecules so tightly that a thin coating of silicone makes a good water repellent.
7. The word "adhere" applies as well, and "adhesion" similarly is used to describe adsorption. "Cohesion," by contrast, refers to a substance's proclivity to cling to itself.

adsorb strongly on materials such as silica gel that also are dipolar. The area of adsorbent available is crucial as well, and since the ratio of the surface area to volume increases the more finely a material is divided (for spheres, the ratio is inversely proportional to the radius), soils having the smallest particles are the best adsorbers of water.

The individual adsorbed molecules may move around on the surface of an absorbent, but wherever located they tend to remain adhered. Water adsorbs (adheres) particularly strongly to many materials because of hydrogen bonding; however, with other substances different forces, including capillary action at pitted or otherwise irregular solid-liquid interfaces, typically determine the degree of adsorption. Thus, in general, adsorption is thought to depend in part on capillary action.

Webster's defines capillarity (capillary action) as: "the action by which the surface of a liquid where it is in contact with a solid (as in a capillary tube) is elevated or depressed depending on the relative attraction of the molecules of the liquid for each other and for those of the solid." Recall that a quantitative measure of the attraction between molecules is the liquid's surface tension. Water's surface tension (in dynes/cm) is 73, which means that the force of attraction pulling the water's surface molecules towards each other amounts to 73 dynes per centimeter length of an arbitrary line drawn on the surface. The force is directed at right angles to the line wherever it is drawn. The effect of this force is readily seen by carefully placing an ordinary sewing needle on a water surface. As illustrated in **Figure A.3**, the needle depresses the surface until its weight W just balances the upward components of the tension forces σ lying in the now-warped surface. If the needle is to sink it must pull apart water molecules, and the surface tension force opposes that.

Surface tension actually is interfacial tension, since when two substances meet at a surface the surface tension depends on the degree of mutual attraction between the molecules on the two sides of the surface. People tend to use the term "surface tension" to mean the interfacial tension when the interface is between a liquid and air. Water's surface tension, i.e. its interfacial tension, when in contact with air is the stated 73 dyne/cm, and the interfacial tension between water and ice is less, about 30 dyne/cm. Water's interfacial tension with soap or with oil is much lower than that between water and air, or water and ice, or water and glass.

When a liquid is placed in an open container, the configuration of the surface between the liquid and the air above is level except near the wall of the container. If the molecules in the liquid are less attracted to those in

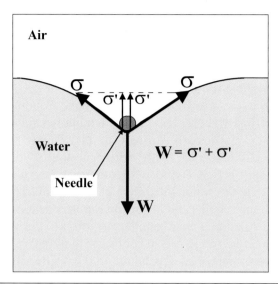

Figure A.3 When a needle is placed on water it presses the surface down until the vertical components of the surface tension force (σ'+ σ') just balance the downward force created by the gravitation force of the earth's attraction on the mass of the needle (W).

the wall than to themselves, the liquid will draw away from the wall to form a downward-curving surface, as shown at left in **Figure A.4**, and if the attraction of the wall molecules is stronger than the self-attraction between the molecules in the liquid the liquid's surface curves upward as shown on the right-hand side of the diagram. (Refer also to Figure 3.5 showing the different behaviors of water and mercury.)

Because it is a definitive measure of the adhesive forces across boundaries and the relevant cohesive forces near them, surface tension (interfacial tension) comes into its own by permitting a simple calculation of the angle at which the surface between a liquid and a gas or between two liquids intersects a solid. This is because the surface at the juncture is stable only if the interfacial tension forces balance. This balance occurs when the interfacial tension between the wall and the vapor $\sigma_{\text{Wall-Vapor}}$ minus the interfacial tension between the wall and the liquid $\sigma_{\text{Wall-Liquid}}$ equals the projection of the surface tension between the liquid and the vapor $\sigma_{\text{Liquid-Vapor}}$ on the wall. This relationship expressed in equation form is

$$\sigma_{\text{Wall-Vapor}} - \sigma_{\text{Wall-Liquid}} = \sigma_{\text{Liquid-Vapor}} \cos \alpha.$$

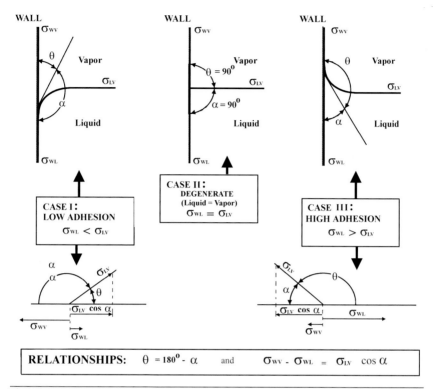

RELATIONSHIPS: $\theta = 180° - \alpha$ and $\sigma_{WV} - \sigma_{WL} = \sigma_{LV} \cos \alpha$

Figure A.4 The upper left and lower left parts of the diagram schematically describe the situation involving mercury and glass where the attraction between the glass and the mercury is almost nonexistent (Case I). At the wall the liquid clings to itself more and therefore it curves downward, and so the contact angle α is very high, greater than 90 degrees. Case II is degenerate. Case III, at right, depicts the situation involving water. Here, the contact angle is low and the liquid curves up the wall because its molecules have high affinity for the molecules in the wall.

The angle α is called the ***contact angle***, and as shown in Figure A.4, $\alpha = 180° - \theta$, where θ is the angle at which the surface meets the wall.[8] Low or near-zero values of the contact angle α correspond to extremely high adhesion of the liquid on the wall, that is, complete wetting such as occurs between water and glass. The interfacial tension is high, as is the interfacial free energy. It is the energy that would be required to break the

8. This relationship and the geometry shown in Figure A.4 also holds for two liquids (designated as 1 and 2) in contact with a wall (W), and then the more general form of the equation is $\sigma_{W1} - \sigma_{W2} = \sigma_{1,2} \cos \alpha$.

surface. The equivalent quantities, interfacial tension and free energy, decrease with increasing α, until in the limit α = 180° the interfacial tension and interfacial free energy become zero. That allows the liquid to separate from the wall spontaneously, i.e., without any work having to be applied from outside. With its near-zero contact angle, water always leaves a film on glass, but it can be shaken from paraffin (α ~ 110°). Mercury is a heavy substance with very high interfacial tension with air, 470 dynes/cm, but little affinity for glass, as is expressed by its high contact angle, α ~ 140°.

Water placed in a glass container with vertical walls climbs up the walls a short distance, but if a glass tube (a capillary) is placed in the container, the water will rise higher within the tube, as was shown earlier in Figure 3.5. The rise is to a level where the total upward force due to interfacial tension balances the downward force due to the weight of the water in the column. Since for a water-glass interface the contact angle α is nearly zero, the upward force is essentially equal to the surface tension force σ per unit length times the length of the line of contact around a capillary of radius r. Thus the total upward force is $2\pi r\sigma$. The downward force is equal to (volume of the column) × (density of liquid contained ρ) × (acceleration of gravity g). The volume is the area of the column πr^2 times the height h. Thus:

$$2\pi r\sigma \text{ (upward force)} = \pi r^2 h \rho g \text{ (downward force)}$$

and by rearranging terms we find $h = 2\sigma/r\rho g$, the result previously derived in Chapter 3 where the matter of contact angle was glossed over without mention.

The acceleration of gravity g is a fixed quantity, and for a given wall material and liquid, so are the surface tension σ and density ρ, thus the height the liquid rises in the capillary depends strictly on the radius of the tube: the smaller the tube the higher the rise. If the contact angle α exceeds 90°, the liquid in the capillary falls rather than rises, and the meniscus becomes convex rather than concave, as illustrated in Figure 3.5.

▪ Derivation of Various Relationships among Pressure, Temperature, and Free Energy Pertaining to the Coexistence of Ice and Water

The total free energy of a system G per unit mass is defined as the system's internal energy E plus the product of its specific volume (volume per unit mass, not total volume) V and pressure P minus the product of its temperature T and a quantity s, called entropy, that is a measure of how much energy is locked up in a system's molecular arrangements and therefore unavailable. In equation form:

$$\text{Absolute Free Energy} = G = E + PV - Ts$$

Specification of the magnitude of free energy requires reference to a level that is unknown because the absolute magnitude of the entropy s cannot be determined. This is not a significant problem because knowledge of how a system will behave in a particular circumstance requires only information about differences in free energy. A specific example, one of particular interest here, is the difference in free energy between water in the liquid and the solid state at a given temperature. Another relevant question is: How does the free energy of liquid and solid water vary with change in temperature and pressure, and what are the consequences of the variation? We begin by seeking to determine the change in free energy ΔG when the temperature changes by ΔT and the pressure by ΔP. The first step is to alter the above equation by adding the increments ΔG, ΔP, and ΔT in the appropriate places, thereby obtaining the expression:

$$\Delta G + G = E + V(P + \Delta P) - (T + \Delta T)s = E + VP + V\Delta P - Ts - \Delta Ts.$$

Then, by subtracting the first equation from the second, we obtain:

$$\text{Free Energy } \Delta G = V\Delta P - \Delta Ts, \qquad (1)$$

and now call ΔG the free energy (chemists also call it chemical potential), with the understanding that it is only the free energy relative to some unknown level. This simple equation leads to some powerful results when applied to ice and water. For example, we know when water freezes it does so because at that temperature the free energy of ice ΔG_{ice} must be exactly equal to that of water ΔG_{water}. Therefore:

$$\Delta G_{ice} = \Delta G_{water} = V_{ice}\Delta P_{ice} - \Delta Ts_{ice} = V_{water}\Delta P_{water} - \Delta Ts_{water}.$$

By rearranging terms we have,

$$\Delta T(s_{water} - s_{ice}) = V_{water}\Delta P_{water} - V_{ice}\Delta P_{ice}.$$

Fortunately, a simple relationship exists between the entropy of water s_{water} and the entropy of ice s_{ice}. The relationship comes directly from the definition of entropy, which is $s = Q/T$ where Q is the quantity of energy transferred into or out of a system at temperature T when some action involving heat transfer or mechanical work occurs. When water freezes into ice at some absolute temperature T_o the energy given off is the heat of fusion L. Therefore $s_{water} - s_{ice} = L/T_o$. Substituting this relationship into the equation yields one form of the fundamental expression known as the Clausius-Clapeyron equation,

$$\Delta T = (V_{water}\Delta P_{water} - V_{ice}\Delta P_{ice})T_o/L. \tag{2}$$

Substituting $\Delta P_{water} = P_{water} - P_o$ and $\Delta P_{ice} = P_{ice} - P_o$ into the expression and then setting, $T_o = 0°C$ and $P_o = 0$ at 1 Atm yields a general expression which is another variant of the Clausius-Clapeyron equation,

$$T - T_o = (V_{water}P_{water} - V_{ice}P_{ice})T_o/L \tag{3}$$

which indicates that the pressure on the ice is greater than on the water at temperatures below 0°C.

However if on Equation 2 we impose the condition that the pressure on the ice is the same as that on the water, we then obtain an expression for the change in freezing point with pressure:

$$\Delta T = (V_{water} - V_{ice})\Delta P T_o/L \tag{4}$$

Recall that V_{water} and V_{ice} are specific volumes, so at a pressure of 1 Atm and temperature $T_o = 0°C = 273$ K, we know that V_{ice} is about 9 % greater than V_{water}. Thus (T is negative, and so the equation tells us that the freezing point of water decreases with increasing pressure. The decrease is small, amounting to only about 0.0074°C per Atm.

Another very important result comes from Equation 2 if we ask what happens when we allow the pressure on the water to be below that on the ice while requiring that the ice remain at constant pressure, i.e., $\Delta P_{ice} = 0$. Then

$$\Delta T = (V_{water}\Delta P_{water} - V_{ice}\Delta P_{ice})T_o/L = V_{water}\Delta P_{water} T_o/L$$

and solving for ΔP_{water} we get

$$\Delta P_{water} = \Delta T L/T_o V_{water}.$$

Setting $\Delta P_{water} = P_{ice} - P_{water}$ and $\Delta T = T_o - T$ we have an expression for variation of the pressure difference between ice and water with temperature

$$P_{ice} - P_{water} = (T_o - T)L/T_o V_{water}. \qquad (5)$$

▪ Discussion of the Derived Relationships

Equation 4 above describes the change in freezing point ΔT when the pressure changes by some amount ΔP:

$$\Delta T = (V_{water} - V_{ice})(PT_o/L$$

In this expression V_{water} and V_{ice} are the specific volumes of water and ice, respectively; that is, the volume of each per unit mass. The quantity L is the heat of fusion of ice, and T_o (in degrees Kelvin) is the normal freezing temperature at atmospheric pressure (0°C = 273.15 K). This expression shows that the change in freezing temperature is directly proportional to the change in pressure, and since $V_{water} - V_{ice}$ is negative (because V_{ice} is 9% greater than V_{water}) the freezing temperature falls as the pressure increases. Water is almost unique in this regard, since most substances contract upon freezing, forcing $V_{water} - V_{ice}$ to take on positive values which causes the freezing point to rise with increasing pressure. Water's freezing point falls only slightly with pressure, by about 0.0074°C per atmosphere of pressure increase. Therefore the pressure due to the weight of a few meters of overburden depresses the freezing point only a tiny fraction of 1°C, although material a few thousand meters below the surface experiences a pressure-generated freezing point depression of 1° to 2°C.

Another cause of minor freezing point depression is dissolved salt material in soil water. The dissolved salts slightly lower the free energy of the water and so depress the freezing point, typically by about 0.1°C.

Figure A.5 contains a schematic representation illustrating how both increased pressure and salt content depress the freezing point. The top part of Figure A.5 is a standard pressure versus temperature diagram for water that portrays the phase of water substance for any combination of pressure and temperature, whether the phase be solid, liquid, or vapor. If a pressure-temperature combination falls on one of the lines of the diagram, two phases can coexist, those to either side of the line. The diagram shows that ice, water, and vapor can coexist only at one point,

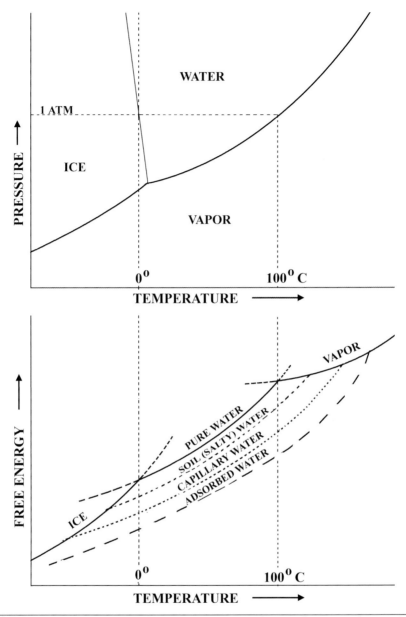

Figure A.5 Top: A conventional pressure-temperature diagram for water substance, its solid lines defining the temperature and pressure conditions at which water substance can coexist in solid, liquid and vapor forms. If the pressure is 1 atmosphere, the melting and boiling points of water are 0°C and 100°C, respectively. A triple point pertains just above 0°C where ice, water and vapor can coexist. Bottom: A free energy-temperature diagram schematically showing how the freezing and boiling points move away from each other with increasing salinity or suction. As the temperature falls, the last water to turn to ice is the adsorbed water because it has low free energy by virtue of being held tightly to soil particles. Based in part on Figure 2-3 of Knight (1967).

the so-called triple point where the three lines intersect.[9] The bottom part of Figure A.5 is a free-energy versus temperature diagram properly aligned with the top part to show how the two types of diagrams relate. Notice that at a pressure of 1 Atm and temperature 100°C the free energies of pure (bulk) water and vapor are equal, and that at pressure 1 Atm and temperature 0°C the free energies of ice and pure water are equal. Also, raising the pressure above 1 Atm shifts where the curves intersect on the lower diagram, to lower freezing temperature and higher boiling temperature. Thus the high pressure created by a skater's blades gliding on the ice enhances the already present thin sheath of premelted water, and a pressure cooker raises the temperature of boiling water to better kill bacteria when preserving foods.

In the lower diagram the dotted-line curve labeled "soil water" schematically portrays the lowered free energy of water containing salts, and the curves below it similarly represent capillary water and adsorbed water. Notice that the lower the free energy of the water represented, the lower falls the freezing point and the higher climbs the boiling point.

---

9. The triple point relates to pure water, water vapor, and ice, but the melting point of ice relates to water saturated with air, and is lower by 0.01°C at pressure 1 Atm. Of that difference the dissolved air accounts for 0.0025°C.

Appendix B
Classification Schemes for Soils

■ Soil Horizons

Within soil that has undergone weathering several distinct levels called horizons may be evident:[1]

> **A Horizon**—The top zone of intense weathering, typically containing organic materials.

> **B Horizon**—The layer just below the A horizon that is moderately weathered and may contain materials leached from the overlying A horizon.

> **C Horizon**—The lowest layer of weathering; within it the weathering is slight.

> **D Horizon**—Underlying unweathered material.

Sublayers within the top horizons may be distinguishable, and are designated by subscripts; for example A_l, A_f or B_h, where *l* means litter, *f* fermented and *h* humic.

---

1. Gieseking, J. E. (1965).

▪ Soil Classification by Type

The nature of the soil within soil horizons depends much on climate, as well as other factors such as vegetation, slope and age. A broad classification scheme of soil types is as follows:

Podzolic—Soils typical of cool, humid, forested regions; usually acidic and dark colored; the B horizon contains aluminum and iron compounds leached from the A horizon.

Lateritic—Red or yellow soils typical of warm climates where aluminum and iron compounds remain in the A horizon.

Pedocals—Soils typical of areas with low rainfall and sparse grass and shrub coverage. May contain saline compounds, have high alkalinity, and low content of organic materials.

Prairie and Chernozem—Soils that develop in areas with intermediate rainfall and grass cover. They are fertile, deep, black soils nearly neutral in pH, and they are highly productive.

Additionally, scientists working with cold-climate soils in different countries have developed several detailed schemes specifically relating to these soils.[2] One classification developed in the old Soviet Union groups the soils into "arctic soils" and "tundra soils," and then within those groups categorizes them mainly according to where they are located because that largely determines the soil drainage. Another more complex scheme adopted by the U.S. Department of Agriculture contains 15 category names not easily remembered that help sort out the soils according to organic content, layer thickness, color, and extent of formation of distinct soil horizons. Perhaps most meaningful to the nonspecialist is the Canadian scheme that breaks the soils into two main groups: *turbic cryosols*, meaning cold soils cryoturbated (stirred up by freeze and thaw processes), and *static cryosols*, meaning cold soils not cryoturbated. This grouping makes an important statement about one of the consequences of repeated freezing and thawing in the active layer: a churning that can destroy the identity of the soil horizons or prevent their formation. See **Table B.1**.

---

2. Washburn (1980)

Table B.1 Canadian Classification of Cold Soils (Cryosolic Order)

TURBIC CRYOSOLS Mineral soils strongly cryoturbated and generally associated with patterned ground		STATIC CRYOSOLS Mineral soils without strong cryoturbations	
Orthic← Turbic Cryosols	Soils strongly cryoturbated with tongues of inter-mixed mineral and organic material, imperfectly to moderately well drained (equivalent to upland tundra and forest tundra)	Orthic Static Cryosols	Soils having a gleyed Bm horizon above the perma-frost table; that is, a sticky clay formed under very wet conditions
Bunisolic↑ Turbic Cryosols	Soils less affected by cryo-turbation and having unbroken Bm horizon similar to the orthics	Brunsolic Static Cryosols	Similar to orthic sub-group but with thicker Bm horizons (equivalent to Arctic Brown Soils in Great Soil Groups of Zonal Classification)
Regosolic→ Turbic Cryosols	Soils lacking Bmy or Bm horizons; surface organic horizon may be present but organic intrusions are lacking in subsoil (generally in high arctic or alpine regions)	Regosolic Static Cryosols	Except for upper part of A horizon, soils without developed soil horizons
Gleysolic↓ Turbic Cryosols	poorly drained soils with mottles and low chromas, and Cg or Bg horizons at mineral surface		

← Orthic—well-developed soils;
↑ Brunosolic—deep dark soils developed from loess;
→ Regosolic—containing much unweathered mineral material;
↓ Gleysolic—with sticky clay formed in waterlogged situations.

Appendix C

Accessible Places Where Permafrost and Active Layer Features Can Be Seen

▪ Collapsed Open-System Pingo

On O'Brien Street $\frac{1}{4}$ mile south of intersection with Goldstream Road near Fairbanks, Alaska. On down the road another $\frac{1}{2}$ mile and off to the east $\frac{1}{4}$ mile is a tree-covered open-system pingo with a collapsed central area.

▪ Cryoplanation Terraces

1) Eagle Summit, 174 km north of Fairbanks on the Steese Highway. 2) Along the Top of the World Highway, within a few kilometers of the Alaska-Canada border, especially on the Canadian side.

▪ Cryoturbation Steps

Many hills along highways running through southern Canada and northern continental United States, including: 1) Along Highway 97 within 20 km to either side of Cache Creek, B.C.; 2) On Highway 29, 10 km north of Hudson's Hope, B.C.; 3) Along Interstate 84 from The Dalles, Oregon, eastward to near the Idaho border, particularly good on north-facing slopes in the vicinity of the Weatherby Rest Stop at Mile 335; 4) Along Highway 52 between Kenmare and Minot, North Dakota, near Coulee, on all except south slopes; 5) One slope of Weir Point on the Custer Battlefield, on some of the steepest (usually lower) slopes of the battlefield, and on the east-facing slopes overlooking the Little Bighorn river in southern Montana; 6) Along Interstate 90 in southern Montana on north-facing slopes at Mile 341 and on both north-and south-facing slopes between Miles 258 and 267; 7) Along Interstate 90 in Idaho, east of Kellogg at Exit 54, well-developed large steps on a south-facing slope.

Open-System Pingos

In the Fairbanks area: 1) Farmers Loop Road, just to south of sharp turn in the road located west of the Dog Mushers; on private property, 2) Miller Hill Extension 0.7 miles north of intersection with Miller Hill and Yankovich roads. This private road is not in good condition but cuts through one side of the pingo and so gives as close a view as one could want. Note collapse feature adjacent to the road. Electrical power pole sits atop the pingo.

Palsas

Parks Highway, Alaska, Mile 206.7. Palsas are in small dying lake on the east side of the road. You can walk over to them but exercise extreme caution to avoid plunging through the moss layer into the lake. Denali Highway, Alaska, Mile 40. Large palsa (or pingo) at left was cut by the building of the road and is rapidly thawing.

Rock Glaciers

Richardson Highway, Alaska, Mile 207.7. The rock glacier is a few hundred meters above the road.

Solifluction Lobes

Eagle Summit, 174 km north of Fairbanks on the Steese Highway, especially pronounced in the valley on the western side of the summit, and many others easily seen in this general area.

▪ Sorted (stone) Polygons

1) At Mile 31.8 and Mile 37 on the Denali Highway west of Paxson. 2) General vicinity of Eagle Summit, 174 km north of Fairbanks on the Steese Highway. 3) General vicinity of Caribou Mountain, 158 km north of Livengood, Alaska, on the Dalton Highway.

▪ Thermokarst (thaw) Lakes

1) In the Fairbanks area: Ballaine Lake on the University of Alaska Campus (Farmers Loop Road); pond on left at entry to Tanana Valley Fairgrounds back parking lot. 2) Mentasta Pass area between Glennallen and Tok. 3) Along the Alaska Highway between White River and the Alaska-Canada border including at Miles 1125.5, 1143 and 1152. Most lakes seen in this area are thermokarst lakes.

▪ Tors

1) Castle Rock, 51 km west of Dawson City, Y.T. on the Top of the World Highway. 2) Angel Rocks, approximately 85 km north of Fairbanks on the Chena Hot Springs Road.

Glossary

The field of permafrost is fraught with a plethora of terms, as is typical of any descriptive field of science. As each of these fields mature, its terminology tends to simplify. I have prepared this glossary with the view that simple is best, but for the benefit of the reader who examines some of the older literature I have included here terminology already archaic or likely soon to fall into that classification. To the extent feasible, I have tried to make this glossary compatible with two other more extended ones: the *Glossary of Permafrost and Related Ground-Ice Terms*, Technical Memorandum No. 142 (Ottawa: National Research Council of Canada) 1988, 156 pp, and the International Permafrost Association's *Multi-language Glossary of Permafrost and Related Ground-ice Terms*, compiled and edited by Robert O. van Everdingen (Calgary, Alberta, Canada: The Arctic Institute of North America) 1998, 286 pp. The entries here referenced to van Everdingen (1998) are basically taken from the latter glossary.

Activation energy of nucleation (also called the free-energy barrier). The free energy associated with the growth of a crystal from zero radius to the critical radius, the radius where further growth or dissolution decreases the free energy of the system.

Active layer. The top part of the ground undergoing annual freeze and thaw. Some permafrost authorities restrict the use of this term to describe the layer of annual freeze and thaw only when it is underlain by permafrost (French, 1988); others use the term to mean the layer of seasonal freeze and thaw within the continuous and discontinuous permafrost zones (van Everdingen, 1998), and still others (Chinese and Russian geocryologists) extend the term to include the layer of seasonal freeze and thaw outside the continuous and discontinuous permafrost zones. See also *seasonally frozen layer* and *seasonally thawed layer*. (In this book I lean toward the Chinese and Russians.)

Active-layer failure. A general term referring to several forms of slope failure or failure mechanisms commonly occurring in the active layer overlying permafrost.[1] A more narrow meaning in use is to describe flow-dominated failures in the active layer. Then it is synonymous with *earth flow, skin flow,* and detachment failure.

Active-layer glide (also called block slide, detachment failure, and active-layer detachment slide). Downhill sliding of the active layer as a unit, usually along the typically ice-rich permafrost table. It may develop because of excess pore-water pressure near the bottom of the active layer.

Adsorption. The process by which a substance (the adsorbent) attaches to its walls the molecules of another (the adsorbate). Van der Waals and other electrostatic forces acting at the interfaces between the molecules in the adsorbate with those in the adsorbent cause adsorption, and charge distributions at the interfaces may be involved.

Aggradational ice. New ice layers or ice that enlarges older ice layers within previously frozen ground. This new ice is in consequence of thickening permafrost and derives from water pulled by cryogenic suction into the thickening permafrost from above, below, or the side.

Alpine permafrost (mountain permafrost, high-altitude permafrost). Permafrost existing at high altitudes in middle and low latitude regions.

Associated liquid. One in which the molecules cling together in temporary groupings, as in liquid water.

Atmosphere (Atm). A unit of pressure equal to that of an atmospheric column at sea level. It equals $1.033 \text{ kg/cm}^2 = 14.7 \text{ lbs/in.}^2$ A column of water with height 10.3 m (33.9 ft) or a column of mercury with height 76 cm (29.9 in) has a pressure of 1 Atm. 1 Atm equals 1.013 bar, and 1 bar equals 100 kilopascal or 10^5 newtons/m^2.

---

1. van Everdingen (1998).

Beaded stream. Stream drainage pattern consisting of small pools interconnected by short straight or angled drainage channels. The pools usually form at the intersections of melting ice wedges.

Block fields, block slopes, block streams. Extensive areas where the surface is covered by moderate-sized to large angular blocks of rock: block fields on near-level areas, block slopes on slopes. The block streams are linear and occur on the steepest available slopes or in narrow valleys. Frost wedging, frost heaving, and solifluction are thought to be important causative processes.

Block slide. See *active-layer glide.*

Braking blocks. Large rocks moving slower than surrounding solifluction, tending to impede the flow. See also *ploughing blocks.*

Bulk water (or free water). Water not in confinement.

calorie (written with small "c"). The amount of heat required to raise the temperature of one gram of water by one degree centigrade. Also called the "gram calorie" or "small calorie." If written with a capital "C" Calorie means the "large calorie" or "kilogram calorie" which is 1000 times larger than the small calorie. Food calories are large calories.

Capillarity. The action by which the surface of a liquid where it is in contact with a solid (as in a capillary tube) is elevated or depressed depending on the relative attraction of the molecules of the liquid for each other and for those of the solid.

Closed-system pingos. Pingos formed from a limited water supply trapped by inward freezing *(permafrost aggradation)* of lakes from their boundaries. See also *open-system pingos.*

Contact angle, α. The angle $= 180° - \theta$, where θ is the angle at which the surface bounding a liquid and its gas (or between two liquids) meets the wall of a container. Low or near-zero values of α correspond to extremely high adhesion of the liquid on the wall, that is, complete wetting such as occurs between water and glass.

Coulomb force. See *Electrostatic force.*

Covalent bonding. Very strong kind of chemical bonding wherein atoms are held together primarily by the magnetic forces between paired electrons.

Critical radius. The radius of a crystal where both any further growth or any dissolution decreases the free energy of the system.

Cryogenic. Pertaining to cold, here specifically referring to freezing and thawing of water.

Cryogenic solifluction. *Solifluction* caused by or associated with cryogenic processes such as frost creep, needle ice creep, and gelifluction.

Cryopediment. Found at the foot of mountain or valley slope, a large cryoplanation surface up to several kilometers wide and many kilometers long. See also *nivation, cryoplanation surface, cryoplanation terrace.*

Cryoplanation surface. Flat or nearly flat erosion surface resulting from freeze and thaw processes.

Cryoplanation terrace. On a hilltop or interrupting an upper or middle slope, a steplike or benchlike planar surface created by intense frost wedging associated with melting snowbanks. See also *nivation, cryoplanation surface, cryopediment.*

Cryosuction. A suction in soil that is directly proportional to the drop in ground temperature below the freezing point, 0°C, and which depends on the size and shape of soil particles.

Cryoturbation. Churning of soil by processes related to freezing and thawing, including frost heave and distortions produced by frost sorting and differential freeze and thaw.

Cryoturbations. Deformed soil profiles produced by cryoturbation; also called involutions and periglacial involutions.

Cryoturbation steps (terracettes). On lightly vegetated steep (>20°) hillsides, staircaselike, typically unsorted, steps created by repeated freezing and thawing over the course of years; seen primarily outside the continuous and discontinuous permafrost zones. The steps lie generally transverse to the slope, and have steps and risers typically ranging in size from 30 to 130 cm.

Dansgaard/Oeschger event. Sudden climatic warmings characterized by rapid changes in oxygen isotopic composition lasting hundreds to thousands of years. See also *Heinrich events.*

Debris islands. *Sorted circles* or *sorted polygons* developed on slopes steep enough to give them irregular shapes.

Deflation (of soil). Removal, typically by wind erosion, of silt and sand particles. In barren areas deflation can create stone pavements.

Detachment failure. See *active-layer glide*.

Dielectric constant, ε. When a substance is placed in an electric field it reduces the field; the dielectric constant is the ratio of the reduced electric field to the original electric field in vacuum.

Dipole moment (electric). The magnitude of charge q of a charge pair q^+ and q^- times the distance separating the two charges.

Earth flow. See *active-layer failure* used in the narrow sense of flow-dominated failure.

Earth hummock. A hummock having a core of silty and clayey mineral soil which may show evidence of cryoturbation.[2]

Electron (also called cathode ray and beta ray). The *Handbook of Chemistry and Physics* states, "The electron is a small particle having a negative electrical charge, a small mass, and a small diameter" [about 10^{-12} cm]. An electron carries the smallest known negative electrical charge, often referred to as an elementary charge or as unit charge [4.8×10^{-10} electrostatic units]. Its mass is 9.1×10^{-28} grams. Other properties of an electron include its magnetic moment, its spin and its wavelength. Both observation and quantum mechanics theory dictate that electrons can behave in wavelike fashion and that the wavelength to be assigned is inversely proportional to the mass and speed of the electron [wavelength = h/mv, where h is Planck's constant, 6.6×10^{-27} erg-sec]. Electrons spin on their own axes like tops, and therefore possess two interrelated quantities, angular momentum (also called spin) and magnetic moment. The angular momentum equals $\frac{1}{2}(h/2\pi)$ gram-cm-cm/sec, and the magnetic moment of 9.2848×10^{-24} J T^{-1}(almost exactly) equals the angular momentum times $2e/m_o$. Thus both quantities are fixed.[3] For most purposes it suffices to think of an electron as a point charge or as a little sphere of finite diameter and definite mass, even though the object's partly particle-like, partly wavelike character may mean that (because $E = mc^2$) the electron mass represents only the energy in the electric field within a certain diffusely bounded region somewhat equal to the electron's stated finite diameter, approximately 10^{-12} cm. Thus if a human eye possessed an

---

2. van Everdingen (1998).
3. The magnetic moment of the electrons is considered to be the source of the magnetic behavior of materials such as iron and nickel.

ability to see solid objects as small as electrons, it is not obvious that there would be anything there to see.

Electronegativity. A measure of how strongly atoms grasp their electron swarms. Hydrogen, sodium, and cesium are among the least electronegative atoms, and fluorine is very graspy, having the highest electronegativity of all elements.

Electrostatic force (coulomb force). The inverse square force between electrically charged particles, repulsive for charges of like sign, and attractive for charges of opposite sign.

Epigenetic (permafrost or a permafrost feature). That formed in previously deposited soil.

Free-energy barrier. See *Activation energy of nucleation.*

Frost creep. The ratchetlike downslope transport of particles as the result of frost heaving and subsequent settling upon thawing. (A somewhat unfortunate name, because the usual usage of 'creep' is plastic flow, not up and down motion.)

Frost heave. Expansion in volume of soil due to the formation of segregation ice, most generally expressed as upward movement of the ground surface.

Frost mound. Any mound-shaped landform produced by ground freezing combined with accumulation of ground ice due to groundwater movement or migration of soil moisture.[4]

Frost pull (sometimes called frost jacking). A mechanism believed responsible for lifting stones and other objects out of the ground during frost heaving wherein the freezing ground adheres sufficiently to the sides of the object to lift it as the ground heaves.

Frost push. A mechanism believed responsible for lifting stones or other objects out of the ground through the formation of segregation ice below the object.

Frost wedging (also called frost riving, frost shattering, frost splitting, gelifraction, and congelifraction). The splitting of material, usually rocks,

---

4. van Everdingen (1998).

by the pressure of the freezing of water in cracks, crevices, pores, joints, or bedding planes.

Frozen fringe. The zone in a freezing, frost-susceptible soil between the warmest *isotherm* at which ice exists in pores and the isotherm at which the warmest ice lens is growing, in other words, the zone near the freezing front containing both water and ice.

Gas hydrates (clathrates). Ice-like crystalline solids formed from methane or other gas and water molecules. These may exist in large quantities near the base of permafrost.

Gelifluction. A kind of solifluction involving slow downslope flow of soil weakened by freeze and thaw processes and especially by the development of excess pore-water pressures in the layer just above frozen soil. Gelifluction implies the presence of seasonal frost or permafrost.

Geocryology (earth + cold + study). The study of the frozen or near-frozen parts of the earth. See also *permafrost* and *periglacial.*

Geofabric. A tough synthetic textile laid down in embankments or foundations to give strength and prevent the movement of soil fines upward.

Greenhouse effect. Term applied to the atmosphere because, like a greenhouse, the atmosphere allows short-wave (white) light to enter the earth system but then traps part of the light after it is reradiated at longer wavelength from the ground surface, causing an increase in air temperature.

Heat capacity. The quantity of heat required to raise the temperature of a substance by one degree, usually expressed in calories per gram per degree centigrade. See also *specific heat.*

Heinrich events. Pulses of ice-rafted detritus into the North Atlantic that indicate rapid melting of adjacent continental ice, mostly from the Laurentide Ice Sheet, showing a periodicity of 6000 to 10,000 years (about 8400 years). See also *Dansgaard/Oeschger event.*

High-center polygons. Large unsorted polygons having centers elevated above the boundaries where (often inactive or decaying) ice wedges occur. See also *low-center polygons.*

Hydration shattering. A failure in the rock caused by repeated expansion and contraction of the rock as its pore boundaries absorb and then desorb (lose) water molecules with changes in temperature, a process most effective in fine-grained rocks because of the high surface area of their pores.

Hydraulic conductivity. The measure of a substance's ability to transmit water, specifically, the volume of fluid passing through a unit cross section in unit time under the action of a unit hydraulic potential gradient.

Hydrogen bond. A weak but important chemical bonding force created when the electron of a hydrogen atom approaches an electron pair of another atom.

Ice wedge. Wedge-shaped deposit of foliated ice formed from the repeated freezing of ground water in a fissure opened when the ground contracts in response to rapid and intense lowering of soil temperature. Ice wedges typically form in polygonal arrays.

Ice-wedge cast. The site of a former ice wedge filled in with collapsed soil, usually angular gravel because it preserves the site better than fine-grained soil that is likely to flow.

Interfacial tension. See *surface tension.*

Intrusive ice (also called **injection ice** and **pingo ice**). Ice formed from water under hydrostatic pressure, the ice typically developing in the form of veins, lenses or layers.

Involutions. See *cryoturbations.*

Ionic bonding (also called electrostatic and valence bonding). A strong chemical bonding force caused by the electrostatic attraction between two atoms in consequence of distortions to the electron swarms in a way that makes the attractive force between the atoms stronger than the repulsive force.

Isotherms. Lines on a chart or map connecting locations of equal temperature.

Latent heat of fusion. The amount of energy required to change a substance from the solid to liquid state without changing its temperature; for water, 80 calories/gram.

Latent heat of vaporization. The amount of energy required to change a substance from the liquid to the gaseous states without changing its temperature; for water, 540 calories/gram.

Liquid limit. The condition wherein soil contains so much water that it loses its cohesiveness and begins to flow as a liquid.

Loam. A soil containing more than 35% clay with lesser silt not exceeding 27%. (If the proportion of silt to clay is reversed, the soil is called *silt loam.*)

Loess. Silt and sand (plus minor clay) material 80 to 90% in the size range 0.005 to 0.5 mm thought to be transported by wind but which may also have undergone secondary transport by flowing downslope. Typically, the loess that has undergone secondary transport contains more organic material than the primary eolian loess.

Low-center polygons. Large unsorted polygons having centers below the bounding ridges formed by soil warped upward by active ice wedges. See also *high-center polygons.*

Meniscus. The curved surface between a liquid and a gas or between two liquids of different surface tension when confined in a capillary.

Mud boil. A *nonsorted circle* developed in fine-grained soil.

Mud circle. See *mud boil.*

Mudflow. A flow-dominated episodic downslope movement of saturated soil material. Cryogenic mudflows are usually small, and in permafrost areas they usually develop in consequence of thawing ground ice that has become exposed for one reason or another.

Multiple retrogressive flow. Repeated slope failure that, although flow-dominated, produces arcuate ridges retaining some features of the pre-failure slope relief.

Multiple retrogressive slide. Repeated slide-dominated slumping characterized by its production of large arcuate blocks concave downslope. It may involve both unfrozen and frozen material.

Needle ice. See *pipkrakes.*

Needle ice (pipkrake) creep. A kind of solifluction involving slow ratchet-like downslope transport of the very top of the active layer, the surface material being lifted by needle ice. Compare with *frost creep.*

Neutron. According to the *Handbook of Chemistry and Physics,* a neutron is "a neutral elementary particle of mass number 1. It is believed to be a constituent particle of all nuclei of mass number greater than 1....It produces no detectable primary ionization in its passage through matter, but interacts with matter predominantly by collisions and to a lesser extent, magnetically. Some properties of the neutron are: rest mass 1.00894 mass unit; charge, 0; spin quantum number $\frac{1}{2}$." It has magnetic moment, -1.9125 nuclear magnetons, negative because the magnetic moment is directed opposite to the spin. The diameter is similar to that of the electron and the proton, approximately $1 - 10 \times 10^{-13}$ cm.

Nivation. Enhanced local erosion of a hillside by frost action, solifluction, and the flow of meltwater in rills or sheets, and subsurface, at the edges of and beneath lingering snowdrifts. Nivation is considered important to the initiation and growth of cryoplanation surfaces. See also *nivation hollow.*

Nivation hollow. The transverse notch in the riser created by nivation at the leading edge of a cryoplanation terrace.

Nonpolar (molecules). Those with sufficient symmetry that they are electrically neutral when observed from any nearby point.

Nonsorted circle. A nonsorted circle is a patterned ground form that is equidimensional in several directions, with a dominantly circular outline that lacks a border of stones.[5]

Nonsorted polygon. A nonsorted polygon is a patterned ground form that is equidimensional in several directions, with a dominantly polygonal outline that lacks a border of stones.[6] See also *sorted polygon.*

Nonsorted stripes. Patterned ground with a striped and nonsorted appearance due to parallel strips of vegetation-covered ground and intervening strips of relatively bare ground, oriented down the steepest slope available. See also *sorted stripes.*[7]

Nucleation. The initiation of the transformation of an unstable phase (such as supercooled water) to a more stable phase (such as ice).

Open-system pingos. Pingos having an open conduit that supplies artesian water. (See *pingo; closed-system pingos*).

Osmosis. The one-way movement (diffusion) of a substance through a membrane of some sort, such as a plant or animal cell wall.

Palsa. Peaty ice-cored mound a few meters high having segregation ice in its core, typically found in the discontinuous permafrost zone where light

---

5. van Everdingen (1998).
6. van Everdingen (1998).
7. van Everdingen (1998).

snow or windy conditions allow high temperature gradients within projecting portions of the ground surface.

Patterned ground. Topography exhibiting regular patterns of size ranging from a few centimeters to more than 10 meters, typically expressed as differences in elevation, soil composition and vegetation.

Periglacial (near + glacier). Conditions, processes and landforms associated with cold, nonglacial environments—and their study. See also *permafrost* and *geocryology.*

Permafrost. Ground that has remained at or colder than 0°C for at least two consecutive years, that period chosen to avoid inclusion of seasonal frost that might have survived over a summer. Also the field of study relating to the frozen or near-frozen portions of the earth. Simeon W. Muller (1945) first proposed the term.

Permafrost aggradation. Growth of permafrost by increase in its thickness or areal extent.

Permafrost creep. Slow downslope deformation of perennially frozen soil resulting from the action of gravity on the plastic soil and ice mixture.

Permafrost table. The top of the permafrost layer.

Pingo. Conical or elongated hills or mounds containing massive layers of ice formed under hydrostatic pressure. See also *open-system pingos, closed-system pingos.*

Piping (also stoping). Downward transport of fine-grained soil by subsurface waters flowing through voids in rocky debris deposits.

Pipkrakes (or *needle ice,* also called mushfrost, Kammeise, and shimobashira). Vertically elongated crystals of ice that grow just beneath the ground surface, the ice usually derived from water drawn upward by cryogenic suction. Anyone who has lived in an area where the ground freezes has probably walked on pipkrakes.

Planetary albedo. The fraction of incoming solar radiation that reflects from the earth and its atmosphere, about 34%.

Pleistocene epoch. Dates from the present back approximately 1.8 million years. It is the most recent part of the Quaternary geological period, which had duration about 3 million years.

Ploughing blocks. Large rocks moving downslope faster than surrounding material, their motion enhanced by gelifluction beneath the blocks. See also *braking blocks.*

Plug flow. Slow downslope motion of the upper and middle parts of the active layer caused by melting of segregation ice layers near the bottom of the active layer where it is in contact with permafrost. Plug flow might be thought of as a kind of accelerated *cryogenic solifluction* or as a kind of very slow *active-layer failure.*[8] See also *gelifluction.*

Polar (molecules). Those lacking symmetry such that they produce an electric field at any nearby point. See also *nonpolar.*

Polygon. In the field of geocryology, the term polygon refers to closed, multisided, roughly equidimensional patterned-ground features, bounded by more or less straight sides; some of the sides may be irregular.[9] See also *high-center polygons, low-center polygons, sorted polygon.*

Pore ice. That formed from water already present in soil pores.

Premelting. The formation of a liquid film on a solid in contact with its vapor when the temperature of the solid is below its melting point.

Primary frost heaving. The initial frost heaving that results from the expansion of water brought to the freezing front by cryogenic suction and there freezing to form segregation ice. See also *secondary* and *tertiary frost heaving.*

Protalus rampart. A rocky ridge-like deposit beyond the normal toe of a talus slope that has accumulated as a result of a former or present intervening snow bank.

Proton. The *Handbook of Chemistry and Physics* defines a proton as "a positively charged subatomic particle having a mass of 1.67252×10^{-24} g, slightly less than that of a neutron but about 1836 times greater than that of an electron." The proton's electrical charge is identical in magnitude to the electron's, and its mass is 1836.2 times the electron's. Like the electron, the proton has an intrinsic spin (angular momentum) of $\frac{1}{2}(h/2\pi)$, and associated with that spin is an intrinsic magnetic moment. (2.79 nuclear magnetrons, also 1.4104×10^{-27} erg per gauss). The proton radius is about 1.2×10^{-13} cm.

---

8. Williams and Smith (1989) p 130.
9. van Everdingen (1998).

Regelation. Melting of ice crystals under pressure and then refreezing when the pressure is released, or melting at a location where the temperature is high and refreezing at another location where the temperature is lower.

Rockfall. Rock debris, perhaps loosened by freeze and thaw processes, free-falling, bouncing, or rolling down steep slopes.

Rock glacier. A lobate or tongue-shaped body of debris that moves downslope by deformation of contained interstitial ice and ice lenses. The slope of the toe of an active rock glacier is near the angle of repose of the unfrozen material forming the rock glacier.

Rubble sheet. A deposit more than the half composed of blocky material ranging in size upward from 10 cm. One on level or near-level ground may be called a block field, and those on a slope greater than 10° may be called a block slope.

Rubble stream. A linear array of rubble moving down a steep slope or in a narrow valley.

Sand wedges. Wedge-shaped sand deposits with a vertical fabric that results from sand entering the voids formed by polygonal contraction cracks.

Seasonally frozen layer. In North American usage, the *active layer* in areas without permafrost.

Seasonally thawed layer. In North American usage, the *active layer* in permafrost areas.

Secondary frost heaving. The frost heaving that results from the expansion of water drawn through the freezing front by cryogenic suction and that freezes into layers of segregation ice behind the front.

Segregation ice. Ice formed from water drawn into position by cryogenic suction.

Shut-off pressure (also called maximum heaving pressure). The pressure of a soil that is high enough to prevent the formation of segregation ice, and hence stop frost heaving.

Silt loam. A soil containing more than 35% silt with lesser clay not exceeding 27%. (If the proportion of clay to silt is reversed, the soil is called *loam*.)

Skin flow. See *active-layer failure* used in the narrow sense of flow-dominated failure.

Slopewash. Rainwater or snowmelt that runs across the surface of a slope or beneath it, which may cause erosion by carrying soil in suspension or solution.

Soil. As used in this book, the uppermost layer of material at the earth's surface that is frozen or wherein freeze and thaw processes take place.

Soil freezing characteristic curve. A plot of a soil's moisture content against temperature near and below the freezing point.

Soil moisture characteristic curve. A plot of a soil's moisture content against applied suction.

Soil texture. The size range of particles in soil, also the feel of a soil because the two are related.

Solifluction. Slow downslope movement of the top layer of wet, unfrozen soil. See also *cryogenic solifluction.*

Solifluction benches. Steep-fronted benchlike solifluction deposits up to 15 m thick that generally cut across slopes more or less parallel to the contours.

Solifluction lobes. Lobate solifluction deposits up to 5 m thick, 25 m wide and 150 m long that extend downslope in tonguelike fashion.

Sorted circles. A form of patterned ground that is equidimensional in several directions, with a dominantly circular outline, and a sorted appearance commonly due to a border of stones surrounding a central area of finer material.[10]

Sorted polygons. A form of patterned ground that is equidimensional in several directions, with a dominantly polygonal outline, and a sorted appearance commonly due to a border of stones surrounding a central area of finer material.[11]

Sorted steps (stone garlands). Found only on shallow (5° to 15°) slopes, a type of patterned ground with steplike form involving a downslope border of stones embanking a relatively fine-grained bare ground upslope. See also *cryoturbation steps.*

---

10. van Everdingen (1998).
11. van Everdingen (1998).

Sorted stripes. Patterned ground with a striped and sorted appearance, due to parallel strips of stones intervening strips of finer material, oriented down the steepest slope available. See also ***nonsorted stripes***.[12]

Specific heat. The ratio of its thermal (heat) capacity of a substance to the thermal (heat) capacity of water at 15°C. See also ***heat capacity***.

Specific surface (of soil). The total surface area of the soil particles per unit volume or unit weight.

State (of matter). Generally, the term refers to the form of a substance: solid, liquid or gaseous. Water in the solid state is ice. In quantum mechanics, an electron configuration associated with a particular amount of energy wherein electrons can move without appreciable radiation of energy, also called a stationary state.

Static cryosol. A cold mineral soil not stirred up by freeze and thaw processes (not cryoturbated).

Stoping. See ***piping***.

Stress. The external force per unit area on a soil (or other material) that is resisted by the strength of the soil (or other material) if the soil is in equilibrium with the force. Consider a cross-section of a soil body. The stress perpendicular to that cross-section is called the normal stress, and by convention the normal stress is positive if it is a compressive force, and negative if it acts to pull the soil apart (i.e. is a tensile force). Stress directed in the plane of the cross-section is called shear stress. If a soil lies on a slope it is convenient to consider the cross-section in the plane of the slope. Then the weight of overlying material creates both a normal stress and a shear stress. As the slope increases, the normal (compressive) stress declines, and the shear stress increases. If the shear stress exceeds the soil's strength, downslope creeping or catastrophic sliding ensues. Similarly, a soil under compressive stress may compact, or if under tensile stress it may crack apart (as it does during the formation of ice wedges.)

Subsea permafrost. Material beneath the ocean floor with temperature below 0°C. If it contains ice it is called bonded; if not, unbonded.

Supercooling. The phenomenon of a liquid remaining in the liquid state below the temperature of the freezing (or melting) point.

---

12. van Everdingen (1998).

Surface tension. Tension that appears in a surface between two liquids, a solid and a liquid or a gas, or a liquid and a gas. It can be expressed in terms of force per unit length or in energy per unit area. Also called interfacial tension. The interfacial tension between water and air is approximately 73 dynes/cm, and that between water and ice is approximately 30 dynes/cm.

Syngenetic (**permafrost** or a permafrost feature). That formed concurrently with deposition of soil.

Talik. Unfrozen soil between discontinuous blocks of frozen soil or between the permafrost table and the bottom of the active layer.

Tertiary frost heaving. Frost heaving due to the accumulation of aggradation ice within previously frozen soil.

Thaw bulb (or basin). In permafrost, the ground strata below a house, pipeline, river, lake or other heat source that are thawed.

Thaw consolidation. Reduction in volume of a frozen soil as it thaws; the reduction is caused by 1) the reduction in volume of any ice contained as it liquefies, 2) a reduction in pore size as grains or soil grain aggregates formed during freezing reorient and come closer together during the thaw process, and 3) escape of pore and excess water. If thawing is so rapid that the excess water (from melting of contained ice) cannot escape as rapidly as it forms, then the water pore pressure increases, and that may cause the soil to lose so much strength it slides downslope. The loss of strength occurs because the soil's ability to resist shear (its shear strength) is equal to the sum of soil's cohesion forces plus the effective stress acting perpendicular to the slope (the normal stress), and that effective normal stress is directly proportional to the weight of overlying material minus the pore pressure. Thus high pore pressure can lead to slope failure during thaw consolidation.

Thaw slump. A shear failure wherein blocks of frozen or unfrozen sediments break away, typically along a concave slip face.

Thermokarst. Topographic depressions resulting from the thawing of ground ice.

Thermokarst lakes (thaw lakes). Lakes created by or enlarged by the thawing of ground ice.

Thermomolecular pressure. In the premelting model of segregation ice formation, the quantity $\rho_{ice}L(T_o - T)/T_o$ which acts in addition to the hydrodynamic pressure in the premelted layer to hold ice and soil boundaries apart.

Thermopile. A passive heat pump placed in the ground to prevent thawing of frozen ground or to make the ground freeze.

Thufur. Perennial hummocks formed in either the active layer in permafrost areas or in the seasonally frozen ground (in this book, considered as part of the active layer) in nonpermafrost areas during freezing of the ground. Thufur form in the warmer part of the discontinuous permafrost zone and also under conditions of maritime seasonal frost.[13]

Tor. Rocky, typically vertical or nearly vertically walled rocky remnant left by the weathering processes, in some instances those that generate cryoplanation surfaces.

Translation (in ice crystals). Slipping of ice crystals past others when partial melting occurs.

Transpiration. The process by which water is lost as vapor by movement to the air through membranes or pores of living bodies. Compare with evaporation, the loss of water by direct transfer to the vapor state from a wet surface, also sublimation, the process of water passing directly from the solid to the vapor state.

Transport coefficient. The effective hydraulic conductivity encompassing the movement of water molecules by any process acting, such as regelation.

Turbic cryosol. A cold soil stirred up by freeze and thaw processes (cryoturbated).

Turf hummock (or bog hummock). A hummock consisting of vegetation and organic matter with or without a core of mineral soil or stones.[14]

Van der Waals attraction. A weak electrostatic bonding force effective between atoms or molecules in proximity (as in the liquid state) caused by the greater mutual attraction between one nucleus and another's electron swarm than the sum of all repulsive forces. The larger the atoms or molecules involved, the stronger the Van der Waals attraction.

Viscosity. The internal friction within a liquid or gas that resists change of form; some solids slowly yield and so also exhibit viscosity.

---

13. van Everdingen (1998).
14. van Everdingen (1998).

References

Abbott, Rohn D. (1984) Letter report K-0715 to State Farm Insurance from Shannon & Wilson, Inc. regarding residence at 3 Mile Farmers Loop Road. June 24.

Army (1966) Calculation Methods for Determination of Depths of Freeze and Thaw in Soils—Emergency Construction, Department of the Army Technical Manual TM 5–892–6 (also Department of the Air Force Manual AFM 88–40, Chap. 46) (Washington, D.C.: Departments of the Army and the Air Force) 75 pp.

Balcourt, John (1982) Selecting the right road embankment, *The Northern Engineer* 14, No. 2, 25.

Bard, Edouard (1999) Ice age temperatures and geochemistry, *Science* 284, 14 May, 1133–34.

Barsch, Dietrich (1988) Rockglaciers, in *Advances in Periglacial Geomorphology*, M. J. Clark, Ed. (New York: John Wiley & Sons Ltd.) 69–89.

Battle, W. R. B. (1960) Temperature observations in bergschrunds and their relationship to frost shattering, in W. V. Lewis, Ed., *Norwegian Cirque Glaciers*, London Royal Geog. Soc. Research Series 4, 83–95.

Beuf, Bernard et al. (1971) Les grès du Paléozoic Inférieur au Sahara, *Inst. Français du Pétrole, Science et Technique du Pétrole* 18, Paris, Editions Tech., 464 pp.

Black, R. F. (1974) Ice-wedge polygons in northern Alaska, in *Glacial Geomorphology*, D. R. Coates, Ed. Annual Geomorphology Series, 1974, 5th Proc. (Binghamton, New York: State University of New York, 398 pp) 247–75.

Black, R. F. (1978) Fabrics of ice wedges in central Alaska, *Third International Conference on Permafrost (Edmonton, Alberta, 10–13 July 1978), Proc. Vol 1* (Ottawa: Canada National Research Council, 947 pp) 248–53.

Bogolomov, N. S. and A. N. Sklyarevskaya (1973) On explosion of hydrolaccoliths in southern part of Chitinskaya Oblast, in *Siberian naleds [Naledi Sibiri]* V. R. Alekseyev, Ed, U.S. Army Corps of Engineers, Cold Regions Research and Engineering Laboratory Draft Translation, 399 pp.

Brewer, M. C. (1958) Some results of geothermal investigations of permafrost in northern Alaska, *American Geophysical Union Transactions* 39, No 1, 19–26.

Britton, M. E. (1967) Vegetation of the arctic tundra, in *Arctic Biology*, 2nd Edition, H. P. Hansen, Ed. (Corvallis, Oregon: State University Press) 67–130.

Brooks, C. E. P. (1973) Climate and climatology, *Encylopaedia Britannica*. 5, 927.

Brown, Jerry (1967) An Estimation of the Volume of Ground Ice, Coastal Plain, Northern Alaska, U.S. Army Materiel Command Cold Regions Research and Engineering Laboratory Research Tech. Note, 22 pp.

Brown, R. J. E.(1975) *Permafrost in Canada* (Toronto: University of Toronto Press).

Burgess. M. M., S. E. Grechischev, P. J. Kurfurst, E. S. Melnikov and N. G. Moskalenko (1993) Monitoring of engineering-geological processes along pipeline routes in permafrost terrain in Mackenzie River Valley, Canada and Nadym Area, Russia, *Proceedings, Sixth International Conference on Permafrost, Beijing, China* (July 5–9, 1993) 1, 54–59.

Cailleux, André (1976) Les pingos quaternaires de France, *Rev. Géog.* Montréal, 30, No 4, 374–79.

Calkin, P. E., L. A. Haworth and J. M. Ellis (1987) Rock glaciers of central Brooks Range, Alaska, U.S.A, in *Rock Glaciers*, Giardino, J. R., J. F. Shroder, Jr. and J. D. Vitek, Eds., (Boston: Allen & Unwin) 65–82.

Carson, Charles E., and Keith M. Hussey (1962) Aligned Lakes of Arctic Alaska, *Journal of Geology* 70, No. 4, 417–39.

Chernicoff, Stanley (1995) *Geology* (New York: Worth Publishers, Inc.) 593 pp.

CLIMAP Project Members (1976) The surface of the ice-age earth, *Science* 191, 1131, 19 March.

Coates, Peter A. (1993) *The Trans-Alaska Pipeline Controversy* (Fairbanks, Alaska: University of Alaska Press) 447 pp.

Conner, Billy (1980) Rational Seasonal Load Restrictions and Overload Permits, Report No. FHWA–AK–RD–80–2, (Fairbanks, Alaska: State of Alaska Department of Transportation and Public Facilities) 50 pp.

Crory, F. E., R. M. Isaacs, Edward Penner, F. J. Sanger and J. F. Shook (1984) Designing for frost heave conditions, in *Frost Action and Its Control,* Richard L. Berg and Edmund A. Wright. Eds., (New York: American Society of Civil Engineers) 23–44.

Crowley, T. J. and K.-Y. Kim (1996) Comparison of proxy records of climate change and solar forcing, *Geophysical Res. Lett.* 23, 359.

Cuffey, Kurt M., et al. (1995) Large arctic temperature change at the Wisconsin-Holocene glacial transition, *Science* 270, 455, 20 October.

Dash, J. G., Haiying Fu and J. S. Wettlauffer (1995) The premelting of ice and its environmental consequences, *Rep. Prog. Phys.* 58, 115–67.

Davis, Kenneth S, and John Arthur Day (1961), *Water, the Mirror of Science* (Garden City, New York.: Anchor Books, Doubleday & Company, Inc.) 195 pp.

Davis, Neil (1960) A field report on the Alaska earthquakes of April 7, 1958, *Bull. Seismological Soc. Am.* 50, No. 4, 489–510.

Davis, Neil (1992) *The Aurora Watcher's Handbook* (Fairbanks, Alaska: University of Alaska Press) 230 pp.

Deming, H. G.(1975) *Water, the Fountain of Opportunity* (New York: Oxford University Press).

Dydyshko, P. I., V. G. Kondratyev, M. L. Vasilyev, V. Ya. Prigoda, M. N. Sadakova and A. S. Valuyev, Deformed embankments on mari and the ways of their stabilization, *Proceedings, Sixth International Conference on Permafrost, Beijing, China* (July 5–9, 1993) 1, 155–59.

Edwards, R. Lawrence, et al. (1997) Potactinium-231 dating of carbonates by thermal ionization mass spectrometry: implications for Quaternary climate change, *Science* 276, 782, 2 May.

Ehlig-Economides, Christine (1981) natural gas hydrates—a frozen treasure, *The Northern Engineer* 13, No. 1, 30–35.

Fairbanks Daily New-Miner (1997) "BLM broke the law," page 1, August 26, 1997 and editorial "Hillside or not—it's wetland," page A4, August 27, 1997.

Ferrians, Oscar J. Jr., Rueben Kachadorian and Gordon W. Greene (1969) *Permafrost and Related Engineering Problems in Alaska,* USGS Prof. Paper 678 (Washington D.C.: U.S. Government Printing Office) 37 pp.

Floyd, Peter, (1974) The North Slope Center: how was it built? *The Northern Engineer* 6, No.3, 22.

Fowler, A. C. and C. G. Noon (1997) Differential frost heave in seasonally frozen soils, *Proceedings of the International Symposium on Physics, Chemistry, and*

Ecology of Seasonally Frozen Soils, Fairbanks, Alaska, June 10-12, 1997, Special Report 97–10 (Hanover, New Hampshire.: U.S. Army Cold Regions Research and Engineering Laboratory) 247.

French, H. M. (1977) The pingos of Banks Island, Western Arctic (abs.) *International Union for Quaternary Research (INQUA) Cong., 10th (Birmingham, England, 16–24 August 1977, Abstracts,* 148.

French, H. M. (1988) Active layer processes, in *Advances in Periglacial Geomorphology,* M. J. Clark, Ed. (New York: John Wiley & Sons Ltd.) 151–77.

French, H. M. and J. A. Heginbottom (1983) Eds., *Northern Yukon Territory and Mackenzie Delta, Canada* Guidebook No 3, (Fairbanks, Alaska: Division of Geological and Geophysical Surveys, Department of Natural Resources, State of Alaska) 186 pp.

Friedman, Irving, Carl Benson and Jim Gleason (1991) Isotopic changes during snow metamorphism, in *Stable Isotope Geochemistry: a tribute to Samuel Epstein,* The Geochemical Society, Special Publication No. 3, H. P. Taylor, Jr., J. R. O'Neil and J. R., Kaplan, Eds., 211–21.

Garand, P. (1981) Méthodolgie experimentale permettant l'étude de la gelivite d'un till en fonction du mode de compactage, Mc.Sc.A. Thesis, Univ. Montreal, Dept Civil Eng., 179 pp.

GCASR [Center for Global Change and Arctic System Research] (1995) Preparing for an uncertain future: Impacts of short- and long-term climate change on Alaska, Proceedings of a workshop held in Fairbanks, Alaska.

Gieseking, J. E. (1965) Soil development, *Encyclopaedia Britannica* 20, 926–27.

Goldthwait, R. P. (1976) Frost sorted patterned ground: A review, *Quaternary Research* 6, 27–35.

Goulden, M. L., S. C. Wofsy, J. W. Harden, S. E. Trumbone, P. M. Crill, S. T. Gower, T. Fries, B. C. Daube, S.-M. Fan, D. J. Sutton, A. Bazzaz and J. W. Munger (1998) Sensitivity of boreal forest carbon balance to soil thaw, *Science* 279, 214–17, 9 January.

Grace, John, et al., (1995) Carbon dioxide uptake by an undisturbed tropical rain forest in southwest Amazonia, 1992 to 1993, *Science* 270, 8, 3 November.

Grange, Jack and J. W. Shaw, (1971) Waste treatment in northern Canada, *The Northern Engineer,* Vols 3 & 4, 13–15.

Grave, N. A. (1968) The earth's permafrost beds [Merzlyye tolshchi xemli], Canada Defense Research Board Translation T499R.

Gregory, J. K. et al. (1997), The water dipole moment in water clusters, *Science* 275, 814, 7 February.

Guthrie, Mary Lee (1988) *Blue Babe* (Fairbanks, Alaska: White Mammoth) 31 pp.

Guthrie, R, Dale (1990) *Frozen Fauna of the Mammoth Steppe, The Story of Blue Babe* (Chicago: The University of Chicago Press) 323 pp.

Hays, J. D, (1977) Sensitivity of climate to solar input, UCAR Forum on Solar-Terrestrial Physics (Boulder, Colorado: National Center for Atmospheric Science) 33 pp.

Hays, J. D., John Imbrie and N. J. Shackleton (1976) Variations in the earth's obit: pacemaker of the ice ages, *Science* 194, No. 4270, 1121, 10 December.

Hillel, Daniel (1980) *Fundamentals of Soil Physics* (San Diego, California: Academic Press, Inc.) 413 pp.

Hobbs, P. V. (1954) *Ice Physics* (Oxford: Clarendon Press) 837 pp.

Hopkins, D. M.(1972) The paleogeography and climatic history of Beringia during late Cenozoic time, *Internord*, No 11, 33 pp.

Hughes, Patrick (1976) The year without summer, *EDS* (Environmental Data Service), NOAA, May, 14.

Imbrie, John, and Katherine Palmer Imbrie (1979) *Ice Ages* (Cambridge, Massachusetts: Harvard University Press) 224 pp.

Jahn, Alfred (1975) *Problems of the Periglacial Zone* [Translation published for the National Science Foundation, TT 72-54011] (Warsaw, Poland: Panstwowe Wydawnictwo Naukowe) 221 pp.

Jianheng, Cui, Xu Dongzhou and Chen Hongzhe (1993) Administering countermeasures on embankment thaw settlements of permafrost along the Qinghai Xizang Highway, *Proceedings, Sixth International Conference on Permafrost, Beijing, China* (July 5–9, 1993), 1, 105–10.

Johnson, T. C., E. C. McRoberts and J. F. Nixon (1984) Design implications of subsoil thawing, in *Frost Action and Its Control*, Richard L. Berg and Edmund A. Wright. Eds., (New York: American Society of Civil Engineers.) 45–103.

Johnson-McNichols, Rebecca (1987) Application of stable nitrogen isotope techniques to discern sources of groundwater nitrate in the Fairbanks, Alaska area, *The Northern Engineer* 19, No. 2, 4–11.

Kamensky, R. M., I. P. Konstantinov and V. A. Popov. (1993) Gas pipeline Mastakh-Yakutsk and environment, *Proceedings, Sixth International Conference on Permafrost, Beijing, China* (July 5–9, 1993) 1, 322–25.

Kerr, R. A. (1996) A new dawn for sun-climate links? *Science* 271, 1360, 8 March.

Kerr, R. A. (1998a) The hottest year by a hair, *Science* 279, 315–16, 16 January.

Kerr, R. A. (1998b) Sea floor records reveal interglacial climate cycles, *Science* 279, 1304–05, 27 February.

Knight, Charles A., (1967) *The Freezing of Supercooled Liquids*, (Princeton, New Jersey: D. Van Nostrad Company, Inc.) 145 pp.

Lachenbruch, A. H., et al., (1982) Permafrost, heat flow and the geothermal regime at Prudhoe Bay, Alaska, *Journal of Geophysical Research* 87, B11, 9301.

Lachenbruch, Arthur H. (1957) *Three-dimensional Heat Conduction in Permafrost Beneath Heated Buildings*, USGS Bulletin 1052–B (Washington D.C.: U.S. Government Printing Office) 51–69.

Lacroix, A. V. (1980) A short note on cryoseisms, *Earthquake Notes*, 51, No. 1, January–March.

Lachenbruch, Arthur H. (1962) *Mechanics of Thermal Contraction Cracks and Ice-Wedge Polygons in Permafrost*, special GSA Paper No. 70 (Menlo Park, California: U.S. Geological Survey) 69 pp.

Lachenbruch, Arthur H.(1970*) Some Estimates of the Thermal Effects of a Heated Pipeline in Permafrost*, USGS Circular 632 (Menlo Park, California: U.S. Geological Survey).

Lang, C., M. Leuenberger, J. Schwander and S. Johnson (1999) 16°C rapid temperature variation in Central Greenland 70,000 years ago. *Science* 286, 934–37, 29 Oct.

Lautridou, Jean-Pierre (1988) Recent advances in cryogenic weathering, in *Advances in Periglacial Geomorphology*, M. J. Clark, Ed. (New York: John Wiley & Sons Ltd.) 33–47.

Leffingwell, Earnest de K. (1919) *The Canning River Region Northern Alaska*, USGS Prof. Paper 109 (Washington D.C.: Government Printing Office) 243 pp.

Lewkowicz, Antoni G. (1988), Slope processes, in *Advances in Periglacial Geomorphology*, M. J. Clark, Ed. (New York: John Wiley & Sons) 325–68.

Lotspeich, F. B. (1973) Depth and time of freezing of silty soils, *The Northern Engineer* 5, No. 2, 11–14.

Lunardini, V. J (1993) Permafrost formation time, *Proceedings, Sixth International Conference on Permafrost, Beijing, China* (July 5–9, 1993) 1, 420–25.

Lundqvist, Jan (1969) Earth and ice mounds: a terminological discussion, in *The Periglacial Environment*, Troy L. Péwé, Ed. (Montreal: McGill-Queen's University Press) 203–15.

Mackay, J. R. (1980) The origin of hummocks, western Arctic coast, Canada, *Canadian Journal of Earth Sciences* 17, No 8, 996–1006.

Mackay, J. R. (1981) Active layer slope movement in a continuous permafrost environment, Garry Island, Northwest Territories, Canada, *Canadian Journal of Earth Sciences* 18, No 11, 1666–80.

Mackay, J. R., and W. H. Mathews, (1974a) Needle ice striped ground, *Arctic and Alpine Research* 6, 79–84.

Mackay, J. R., and W. H. Mathews, (1974b) Movement of sorted stripes, the Cinder Cone, Garibaldi Park BC, Canada, *Arctic and Alpine Research*, 6, 347–59.

Mackay, J. R., and J. K. Stager (1966) The structure of some pingos in the Mackenzie Delta area, NWT, *Geog. Bull.* 8, No 4, 360–68.

Mangus, Alfred R. (1986) Thule Air Base, Greenland, foundations on permafrost, *The Northern Engineer* 18, Nos 2 & 3, 51.

McDowall, I. C. (1960) Particle size reduction of clay minerals by freezing and thawing, *New Zealand J. Geol. and Geophys.* 3, No 3, 337–43.

McFadden, T. T. and F. L. Bennett (1991) *Construction in Cold Regions—A Guide for Planners, Engineers, Contractors, and Managers* (New York: John Wiley & Sons, Inc.) 615 pp.

McIntyre, Andrew, and Barbara Molfino (1996) Forcing of Atlantic and subpolar millennial cycles by precession, *Science* 274, 1867, 13 December.

McRoberts, E. C. and N. R. Morgenstern (1974a) The stability of thawing slopes, *Canadian Geotech. J.* 11, 447–69.

McRoberts, E. C. and N. R. Morgenstern (1974b) Stability of slopes in frozen soils, Mackenzie Valley NWT, *Canadian Geotech. J.* 11, 554–73.

Meyer, Bernard S. (1973) Plants and Plant Science, *Encyclopaedia Britannica* 17, 1206–20.

Muller, S. W. (1945) Permafrost or permanently frozen ground and related engineering problems, U.S. Engineers Office, Strategic Engineering Study Special Report 62, 136 pp.(Reprinted 1947, Ann Arbor, Michigan: J. W. Edwards, Inc.).

Mighetto, Lisa and Carla Holmstad (1997) *Engineering in the Far North*, A History of the U. S. Army Engineer District in Alaska (Missoula, Montana: Historical Research Associates, Inc.) 500 pp.

Nelson, F. E., K. M. Hinkel and S. I. Outcalt (1992) Palsa-scale frost mounds, in *Periglacial Geomorphology*, J. C. Dixon and A. D. Abrahams, Eds. (New York: John Wiley & Sons) 305–25.

Ohrai, T. and H. Yamamoto (1985) Growth and migration of ice lenses in partially frozen soil, *Proceedings of the Fourth International Symposium on Ground Freezing, Sapporo, Japan* (Rotterdam: Balkema) 1, 79–84.

Oliphant, J. L., A. R. Tice and Y. Nakano (1985) Water migration due to a temperature gradient in frozen soil. *Proceedings of the Fourth International Conference on Permafrost, Fairbanks, Alaska* (Washington, D.C.: National Academy of Sciences) 951–56.

Oppo, D. W., J. F. McManus and J. L. Cullen (1998) Abrupt climate events 500,000 to 340,000 years ago: evidence from subpolar North Atlantic sediments, *Science* 279, 1335–38, 27 February 1998.

Osterkamp, T. E., and J. P. Gosink (1991) Variations in permafrost thickness in response to changes in paleoclimate, *J. Geophysical Research* 96, No B3, pp 4423–34.

Osterkamp, T. E., and W. D. Harrison (1976) *Subsea permafrost at Prudhoe Bay, Alaska: drilling report and data analysis*, Report UAG R-245, Sea Grant Report No. 76–5 (Fairbanks, Alaska: Geophysical Institute of the University of Alaska) 69 pp.

Overpeck, J. T. (1996) Warm climate surprises, *Science* 271, 1820, 29 March.

Pauling, Linus (1960) *The Nature of the Chemical Bond*, 3rd Ed. (Ithaca, New York: Cornell University Press) 468.

Pauling, Linus (1970) *General Chemistry*, reprint of 3rd Ed. (New York: Dover Publications, Inc.) 959 pp.

Permafrost Subcommittee (1988) *Glossary of Permafrost and Related Ground-Ice Terms*, Technical Memorandum No. 142 (Ottawa: National Research Council of Canada) 156 pp.

Péwé, Troy L.(1975a) *Quaternary Geology of Alaska*, USGS Prof. Paper 935 (Washington, D.C.: U.S. Government Printing Office) 145 pp.

Péwé, Troy L. (1975b) Permafrost: Challenge of the Arctic, *1976 Yearbook of Science and the Future*, Encyclopaedia Brittanica, Inc. 92–105.

Péwé, Troy L. (1977), Ed., *Guidebook to the Quaternary Geology, Central and South-Central Alaska*, Reprint (College, Alaska: Division of Geological and Geophysical Surveys, Department of Natural Resources, State of Alaska) 141 pp.

Péwé, Troy L. (1982) *Geologic Hazards of the Fairbanks Area, Alaska*, Special Report No. 15 (College, Alaska: Alaska Division of Geological and Geophysical Surveys) 109 pp.

Péwé, Troy L. (1997) Personal communication, July 1997.

Péwé, Troy L., G. W. Berger, J. A. Westgate, P. M. Brown and S. W. Leavitt (1997) *Eva Interglaciation Forest Bed, Unglaciated East-Central Alaska: Global Warming 125,000 Years Ago*, Special Paper 319 (Boulder, Colorado: the Geological Society of America) 54 pp.

Péwé, Troy L. and R. D. Reger, Eds. (1983) *Richardson and Glenn Highways, Alaska, Guidebook to Permafrost and Quaternary Geology* (Fairbanks, Alaska: Alaska Department of Natural Resources, Division of Geological and Geophysical Surveys) (reprinted 1993) 263 pp.

Pissart, A. (1974) Determination experimentale des processes responsables des petits sols polygoneaux triés de haut montagne, *Abh. Akad. Wiss Gottingen, ser. 3*, 29, 86–101.

Pissart, A. (1994) The fossil pingos of Wales and the Hautes Fagnes, Belgium, in *Cold Climate Landforms*, David J. A. Evans, Ed. (New York: John Wiley & Sons) 526 pp.

Preisnitz, K. (1988) Cryoplanation, in *Advances in Periglacial Geomorphology*, M. J. Clark, Ed. (New York: John Wiley & Sons) 49–67.

Price, L. W. (1970) Up-heaved blocks: A curious feature of instability in the tundra. *Assoc. American Geographers Proc.* 2, 106–10.

Pringle, Heather (1997), Death in Norse Greenland, *Science* 275, 924, 14 February.

Rabus, Bernhard, Keith Echelmeyer, Dennis Trabant and Carl Benson (1995) Recent changes of McCall Glacier, Alaska, *Annals of Glaciology* 21, 231–39.

Rawlinson, S. F., Editor (1983), *Prudhoe Bay, Alaska, Guidebook to Permafrost and Related Features*, Guidebook No. 5 (Fairbanks, Alaska; Alaska Division of Geological & Geophysical Surveys) 177 pp.

Reger, Richard (1995) Personal communication, Fall 1995.

Reger, R. D. and T. L. Péwé (1976) Cryoplanation terraces: Indicators of a permafrost environment, *Quaternary Research* 6, 99–109.

Reimers, Stephen (1980) Drilling and sampling in frozen ground: a few basics, a few problems, *The Northern Engineer* 12, No 2, 13.

Rein, R. G.., Jr., and C. M. Burrous (1980) Laboratory measurements of subsurface displacements during thaw of low-angle slopes of a frost susceptible soil, *Arctic and Alpine Research* 12, 349–58.

Retallack, Gregory J. (1997) Early forest soils and their role in Devonian global change, *Science* 276, 583, 25 April.

Rice, Eb (1973) Northern construction: siting & foundations, *The Northern Engineer* 5, No. 1, 11.

Rice, Eb and Amos "Joe" Alter, (1974) Water supply in the north, *The Northern Engineer* 6, No. 2, 10.

Sackinger, W. M. (1980) Corrosion of a chilled gas pipeline buried in discontinuous permafrost, in *Annual Report 1979–80* (Fairbanks, Alaska: Geophysical Institute, University of Alaska) 158–59.

Schell, Don (1998) Fossil food, in *Alaska Science Nuggets*, 3rd Printing, Neil Davis Ed. (Fairbanks, Alaska: University of Alaska Press) 159–60.

Schmid, Josef (1955) Der Bodenfrost als morphologischer Faktor (Heidelberg: Dr. Alfred Hüthig Verlag) 144 pp.

Seife, Charles (1996) On ice's surface, a dance of molecules, *Science* 274, 2012, 20 December.

Sellmann, P. V. (1967) Geology of the USA CRREL permafrost tunnel, Fairbanks, Alaska, Technical Report 199 (Hanover, New Hampshire: U.S. Army Materiel Command Cold Regions Research and Engineering Laboratory) 22 pp.

Sellmann, P. V. (1972) Geology and properties of materials exposed in the USA CRREL permafrost tunnel, Special Report 177 (Hanover, New Hampshire: U.S. Army Materiel Command Cold Regions Research and Engineering Laboratory) 16 pp.

Sellmann, P. V., Jerry Brown. R. I. Lewellen, H. L. McKim and C. J. Merry (1975) The classification and geomorphic implications of thaw lakes on the Arctic Coastal Plain, Alaska, Research Report 344 (Hanover, New Hampshire: U.S. Army Materiel Command Cold Regions Research and Engineering Laboratory) 21 pp.

Seppälä, Matti (1988) Palsas and related forms, in *Advances in Periglacial Geomorphology*, M. J. Clark, Ed. (New York: John Wiley & Sons) 247–78.

Seppälä, Matti (1995) How to make a palsa: possibilities of experimental geomorphology, a lecture given at the University of Alaska Fairbanks Museum on September 15, 1995.

Severinghaus, J. P. and E. J. Brook (1999) Abrupt climate change at the end of the last glacial period inferred from trapped air in polar ice, *Science* 286, 930–34, 29 October.

Sharp, R. P. (1942a) Soil structures in the St. Elias Range, Yukon Territory, *J. Geomorphology* 5, 274–301.

Smith, M. W. (1985a) Models of soil freezing, in *Field and Theory, Lectures in Geocryology*, Michael Church and Olav Slaymaker, Eds. (Vancouver, British Columbia: The University of British Columbia).

Smith, M. W. (1985b) Observations of soil freezing and frost heaving at Inuvik, Northwest Territories, Canada, *Canadian Journal of Earth Sciences* 22, No 3, 283–90.

Stager, J. K.(1956) Progress report on the analysis and distribution of pingos east of the Mackenzie Delta, *Canadian Geographer* 7, 13–20.

Stella, Damien F. (1986) A qualitative approach to minimizing differential frost heave, *The Northern Engineer* 18, No 4, 10.

Strock, Clifford and R. L. Korel (1959) Eds., *Handbook of Air Conditioning, Heating and Ventilating*, 2nd Ed. (New York: The Industrial Press).

Strömquist, Lennart (1973) Geomorfologiska studier av blockhav och blockfält i norra Skandinavien [Geomorphological studies of block-fields in northern Scandinavia], Uppsala Univ., Naturgegrafiska Inst., Avdelningen for Naturgeografi, UNGI Rapport 22, 161 pp.

Sturm, Matthew, Jon Holmgren and G. E. Liston (1995) A seasonal snow cover classification system for local to global applications, *Journal of Climate* 8, No 5, Part II, 1261–83.

Taber, S. (1929) Frost heaving, *Journal of Geology* 37, No. 5, 428–61.

Thorn, Colin (1988) Nivation: A geomorphic chimera, in *Advances in Periglacial Geomorphology*, M. J. Clark, Ed. (New York: John Wiley & Sons) 3–31.

Trabant, Dennis, and Carl Benson (1972) Field experiments on the development of depth hoar, in *Studies in Mineralogy and Precambrian Geology*, Memoir 135, B. R. Doe and D. K. Smith, Eds. (Boulder, Colorado: The Geological Society of America) 309–21.

Tricart, Jean (1970) *Geomorpholgy of Cold Environments* [Géomorphologie des Régions Froides], translated by Edward Watson (London: Macmillan; New York: St. Martin's Press) 320 pp.

Vandenberghe, J. (1988) Cryoturbations, in *Advances in Periglacial Geomorphology*, M. J. Clark, Ed. (New York: John Wiley & Sons) 179–97.

van Everdingen, Robert O. (1998) editor, International Permafrost Association's *Multi-language Glossary of Permafrost and Related Ground-ice Terms* (Calgary, Alberta, Canada: The Arctic Institute of North America) 286 pp.

Vitek, J. D. and J. R. Giardino (1987) Rock Glaciers: a review of the knowledge base, in *Rock Glaciers*, Giardino, J. R., J. F. Shroder, Jr. And J. D. Vitek, eds., (Boston: Allen & Unwin 1–26.

Wahrhaftig, Clyde (1949) The frost rubbles of Jumbo Dome and their significance in the Pleistocene chronology of Alaska, *J. Geol.* 57, 216–31.

Wahrhaftig, Clyde, and Allan Cox (1959) Rock glaciers in the Alaska Range, *Geol. Soc. Am. Bull.* 70, 383–436.

Washburn, A. L. (1973) *Periglacial Processes and Environments* (London: Edward Arnold Ltd.) 320 pp.

Washburn, A. L. (1980) *Geocryology* (New York: John Wiley & Sons) 406 pp.

Weniger, Willibald, (1940) *Fundamentals of College Physics* (New York: American Book Company) 323.

Wernecke, L. (1932) Glaciation, depth of frost, and ice veins of Keno Hill, Yukon Territory, *Eng. and Mining Journal* 133, 38–43.

Werner, B. T. (1999) Complexity in natural landform patterns, *Science* 284, 102–04, 2 April.

Wescott, E. M., S.-I. Akasofu and W. M. Sackinger (1979) Induced currents in the trans-Alaska pipeline, in *Annual Report 1978–79* (Fairbanks, Alaska: Geophysical Institute, University of Alaska) 138–39.

Wettlaufer, J. S. (1999a) Crystal growth, surface phase transitions and thermonuclear pressure, in *Ice Physics and the Natural Environment*, (NATO ASI Series, Vol. I 56) J. S. Wettlaufer, J. G. Dash and Norbert Untersteiner, Eds. (Berlin: Springer-Verlag) 39–67.

Wettlaufer, J. S. (1999b) Impurity effects in the premelting of ice, *Phys Rev. Lett.* 82, 2516–21.

Wettlaufer, J. S., M. G. Worster, L. A. Wilen and J. G. Dash (1996) A theory of premelting dynamics for all power law forces, *Phy.s Rev. Lett.* 76, 3602–05.

Williams, Peter J. (1999) The freezing of soils: ice in a porous medium and its environmental significance, in *Ice Physics and the Natural Environment*, (NATO ASI Series, Vol. I 56) J. S. Wettlaufer, J. G. Dash and Norbert Untersteiner, Eds. (Berlin: Springer-Verlag) 219–39.

Williams, Peter J. and Michael W. Smith (1989) *The Frozen Earth* (Cambridge: Cambridge University Press) 306 pp.

Williams, Peter J., (1989) *Pipelines & Permafrost*, reprint (Ottawa: Carleton University Press) 129 pp.

Wimer, D. C. (1965) Physical nature and properties of soil, *Encyclopaedia Britannica* 20, 926–27.

Wimmler, N. L.(1926) Notes on the thawing of frozen ground by the Fairbanks Gold Dredging Company, Fairbanks Creek, Alaska, (and related reports written 1926–29), *Publication MR 195–13* (Fairbanks, Alaska: Alaska Division of Geological & Geophysical Surveys) 36 pp.

Worster, M. G. and J. S. Wettlaufer (1999) The fluid mechanics of premelted liquid films, in *Fluid Dynamics at Interfaces*, W. Shyy, Ed. (Cambridge: Cambridge University Press) In press.

Yu, Zicheng and Ulrich Eicher (1998) Abrupt climatic oscillations during the last deglaciation in central North America, *Science* 282, 2235–37, 18 December.

Zemansky, Gil (1975) Wastewater and Alyeska: North of the Yukon River, *The Northern Engineer* 7, No 2, 41.

Zhang, T. and T. E. Osterkamp (1993) Changing climate and permafrost temperatures in the Alaskan arctic, *Proceedings, Sixth International Conference on Permafrost, Beijing, China (July 5–9, 1993)* 1, 783–88.

Zoltai, S. C., C. Tarnocai and W. W. Pettapiece (1978) Age of cryoturbated organic materials in earth hummocks from the Canadian Arctic, *Third International Conference on Permafrost (Edmonton, Alberta, 10–13 July 1978)* (Ottawa: Canada National Research Council) 325–31.

Index